Commercial
RABBIT RAISING

R. B. Casady, P. B. Sawin,

and J. Van Dam

UNITED STATES DEPARTMENT
OF AGRICULTURE

Fredonia Books
Amsterdam, The Netherlands

Commercial Rabbit Raising

by
R. B. Casady
P. B. Sawin
J. Van Dam

for
United States Department of Agriculture

ISBN: 1-4101-0767-1

Reprinted from the 1971 edition

Fredonia Books
Amsterdam, The Netherlands
http://www.fredoniabooks.com

Contents

Commercial

RABBIT RAISING

R. B. Casady, P. B. Sawin, and J. Van Dam [1]

INTRODUCTION

Americans eat 25 to 30 million pounds of domestic rabbit meat each year. The rabbits come from small rabbitries with three or four hutches and from large commercial producers. Rabbit raising lends itself to both types of production. Rabbit meat is pearly white, fine-grained, palatable, and nutritious. It is a convenient source of high-quality protein and is low in fat and caloric content.

Rabbitskins also have some commercial value. Better grades of rabbitskins may be dressed, dyed, sheared, and made into fur garments and trimmings. Some skins are used for slipper and glove linings, for toys, and in making felt. Fine shreds of the flesh part of the dried skins, which are often left after separating the fur for making felt, are used for making glue. Because of the relatively low value of skins from meat rabbits, a large

[1] Dr. Casady was formerly with the Sheep and Fur Animal Research Branch, Animal Science Research Division, Agricultural Research Service.

Dr. Sawin is responsible for the section on Systems of Breeding; he was with the Roscoe B. Jackson Memorial Laboratory, Hamilton Station, Bar Harbor, Maine, and is now retired.

Mr. Van Dam is responsible for the section on Economics of Rabbit Production; he is farm adviser, Los Angeles County, University of California Agricultural Extension Service.

volume is necessary to market them satisfactorily.

An increasing demand for rabbits for laboratory and biological purposes offers opportunities to breeders living near medical schools, hospitals, and laboratories. Rabbits have made large contributions to research in venereal disease, cardiac surgery, hypertension, and virology, and are important tools in pregnancy diagnosis, infectious disease research, the development of hyperimmune sera, development of toxins and antitoxins, and the teaching of anatomy and physiology. A recent development in the rabbit industry has been the increased use by scientific personnel of various rabbit organs and tissues in specialized research. The availability of these byproducts has greatly facilitated many basic research programs.

The recommendations in this bulletin are based largely on studies at the U.S. Rabbit Experiment Station formerly maintained at Fontana, Calif., by the Sheep and Fur Animal Research Branch, Animal Science Research Division, Agricultural Research Service. At this station, improved methods were developed for producing rabbits for meat, fur, and wool of fine quality, for insuring sanitary surroundings, and for preventing outbreaks of parasitic and other diseases.

This handbook is being issued to help county agricultural agents, State colleges of agriculture, and the U.S. Department of Agriculture answer the many thousands of requests received each year from commercial and professional rabbit producers for information about rabbits.

CHOOSING A BREED

Whether you raise rabbits for meat and fur, wool, laboratory animals, or show stock—select the breeds best adapted to the purpose.

The American Rabbit Breeders Association lists standards for 28 different breeds, and approximately 77 varieties of these breeds of rabbits, to cover characteristics such as type, color, and size; disqualifications also are listed. Table 1 lists some common breeds of rabbits.

Mature animals of the smaller breeds weigh 3 to 4 pounds each; those of the medium breeds, 9 to 12 pounds; and those of the larger breeds 14 to 16 pounds. They also vary widely in color.

Rabbits best suited in size and conformation for producing meat and fur are such medium and large

TABLE 1.—*Some common breeds of rabbits*

Breed	Color	Approximate mature weight	Principal uses
		Pounds	
American Chinchilla.	Resembles the true chinchilla, (*Chinchilla laniger*).	9–12	Show and fur.
Californian	White body; colored nose, ears, feet, and tail.	8–10½	Show and meat.
Champagne d'Argent.	Undercolor a dark slate blue; surface color a blue-white or silver with a liberal sprinkling of long black guard hairs.	9–12	Show and meat.
Checkered Giant	White with black spots on cheek, sides of body, and on hindquarters; wide spine stripe; black ears and nose with black circles around the eye.	11 or over	Show and fur.
Dutch	Black, blue, chocolate, tortoise, steel gray, and gray; white saddle, or band over the shoulder carrying down under the neck and over the front legs and hind feet.	3½–5½	Show and laboratory.
English Spot	Basic body color white; colors of spots: black, blue, chocolate, tortoise, steel gray, lilac, and gray; nose, ears, and eye circles and cheek spots; spine stripe from base of ears to end of tail; side spots from base of ears to middle of hindquarter.	5–8	Show, meat, and laboratory.

TABLE 1.—*Some common breeds of rabbits*—Continued

Breed	Color	Approximate mature weight	Principal uses
Flemish Giant___	Steel gray, light gray, sandy, black, blue, white, and fawn. No two colors allowed on solids.	13 or over__	Show and meat.
Himalayan_____	Same as Californian_____	2½–5_____	Show and meat.
New Zealand____	White, red, or black_____	9–12_____	Show, meat, and laboratory.
Polish_____	White, black, or chocolate; ruby-red eyes or blue eyes.	3½_____	Show and laboratory.
Rex_____	Representative of any breed____	7 or over___	Show and fur.
Satins_____	Black, blue, Havana brown, red, chinchilla, copper, Californian, and white.	8–11_____	Show and fur.
Silver Martens___	Black, blue, chocolate, or sable, with silver-tipped guard hairs.	6½–9½_____	Show and fur.

breeds as the New Zealand, Californian, Champagne d'Argent, Chinchilla and Flemish Giant. White breeds of rabbits, such as the New Zealand White and Californian, are the most desirable for commercial and fur production because white skins usually bring higher prices. Preference among the white breeds is largely a matter of personal choice. Skins are a byproduct of meat production.

If you raise rabbits for laboratory purposes, check with nearby hospitals, laboratories, and city and county health offices to find out the type, age, and size of animals desired.

BN 21010

FIGURE 1.—Representative breeds of meat rabbits. Left—New Zealand white. Right—Californian.

SELECTING FOUNDATION STOCK

When you use young rabbits for foundation stock, you have an opportunity to become acquainted with them and with their habits before they reach the production stage. An inexperienced producer should begin on a small scale, with 2 or 3 bucks and 20 to 30 does, and expand operations as he gains experience and as market demands justify.

When buying breeding stock, deal directly with reliable breeders. Brokers handling live rabbits seldom are able to vouch for the conditions under which their animals were produced. Reliable breeders stand behind the stock they offer and will give references. National, State, and local rabbit breeders' organizations can furnish names and addresses of breeders from whom you can buy stock.

The essential requirements of good foundation stock are health and vigor, longevity, ability to reproduce, and a body type consistent with ability to produce marketable offspring of the desired quality and size.

SYSTEMS OF BREEDING

In planning a breeding program for rabbits, attention should be given to the concepts which have been shown by long years of study to be relatively constant in domesticated mammals. These concepts are described in a number of textbooks of genetics in greater detail than space will permit here. Such books are in college, university, and most of the large city libraries. A breeder today, who attempts to develop a strain without recourse to such material, would be at a distinct disadvantage. It should be understood that this brief article can be only a summary of the sort of information which a breeder needs.

The first concept is that of the *gene pool.* Any breed or other foundation stock selected for breeding constitutes a pool or group of many, perhaps thousands, of hereditary units, commonly referred to as genes. The genes are specifically located in the chromosomes, very small thread-like bodies found in every cell of the body. In the rabbit there are 22 pairs, and their segregation (one member of each pair going to each egg or sperm) in the production of eggs or sperm, plus the ultimate union of egg and sperm at mating and conception, provides the mechanism for transmission of hereditary characteristics from one generation to the next. It also provides the mechanism which in nature insures sufficient variability for adaptation of the species to minor changes in the environment and for its perpetuation.

The gene pool of the rabbit has been modified in many ways during domestication and by selection to establish the different breeds. This pool, in the rabbits at hand, is the breeder's capital stock, and intelligent breeding depends on knowing as much as possible about that pool. How well does it perpetuate itself? How much variation does it transmit that is either good, bad, or indifferent, particularly with respect to reproductive capacity? How much of it is apparent to the breeder, and how much can be revealed only by breeding experience? In spite of all man knows about genetics and reproduction, nature is still the most successful breeder. If this were not so, we would not have the infinite number and variety of species that exist in the world, many of which are known to have existed for many, many centuries. But even nature slips. Species are known to have been lost as a result of circumstances with which they were unable to cope, and malformed offspring are known to occur sometimes in the wild. Nature's success is essentially due to the size of the gene pools of each species, plus the ruthless elimination of the unfit as they appear. These combine to insure a high proportion of successful individuals, and some individuals adaptable to any ordinary change which may occur in the environment in which they live. Ability to adapt to differing environments is the feature which makes for survival and is the mechanism by which species have evolved.

When man steps in, success or failure of his breeding system depends on the genes maintained in the pool and his ability to select those genes intelligently. The first task of the breeder becomes one of devising methods of ascertaining the sort of genes his animals possess and the second is the elimination of undesirables. To accomplish these objectives there are tools at his disposal, the use of which must be clearly understood. Such tools are: selection, outbreeding or outcrossing, and inbreeding.

Selection has been called the keystone of the arch of animal breeding. It has been practiced in the wild since the beginning of life on this earth. In free-roaming animals, such as the rabbit, where the chances of mating between closely related animals are small, unfavorable recessive genes seemingly are rare. Actually, there are many but they tend to be covered up by dominant favorable genes and by their interrelations in the gene pool. Over a long period of time, a wild population continually mating in this way appears to achieve a relatively high degree of homeostasis, or stability, in a variable environment, with seemingly a minimum of variation. When man steps in with artificial selection under domestication, and an artificial small environment the chances of unfavorable recessive genes coming together in any one mating are greatly increased. In standardbred strains, selection over a long period of years by one breeder under one type of breeding may also lead to homeostasis, but when such a strain is put in inexperienced hands, or under a different system of breeding, it may not produce the same results. Selection, although in itself something of a breeding system without the ruthless objectivity found in the wild, becomes highly dependent upon other factors only controllable by the skill and understanding of the breeder. Two factors are of major importance. First is the quality of the gene pool when selection is first started. It is obviously impossible to select for a characteristic, such as high performance, if the genes for this characteristic are not there in the first place. Second is a good environment which will allow the results of the genetic selection to be fully expressed. Feed, housing, and managerial practices are most important. Overfeeding and pampering, however, may cover up poor genes and thus not lead to permanent improvement or stability.

The supports of the keystone at the two outer extremes of the breeding arch are outbreeding and inbreeding.

Outbreeding, or the mating of unrelated rabbits, differs from nature's usual procedure in no way except in the degree of selection.

In the wild, natural selection occurs through the survival of the fittest based upon function in the environment available. Selection by the rabbit breeder, if done intelligently, often proves superior in many ways. Outbreeding with careful selection is generally accepted as a satisfactory procedure for commercial purposes and, with due attention to reproductive capacity, accounts for a moderate degree of improvement of breeds, particularly when carried out according to the breed standards. It cannot produce permanently a high degree of uniformity, even in the hands of the most skilled breeder, nor can it lead to establishment of an outstanding strain with recognizably uniform dependable improvement.

Outcrossing, or hybridization, consists of wide matings between unrelated rabbits. It is usually done between breeds for special purposes. It is the initial step in the establishment of new breeds, because by bringing together a maximum number of unlike genes of the two breeds (or gene pools), a maximum range of variation from which to select is provided. Because of the extreme degree of relationship, such matings in the first one or two generations frequently manifest a maximum amount of fertility, vigor, and growth, commonly referred to as hybrid vigor or heterosis. The first generation is thus often a highly desirable commercial animal especially for meat production. Later generations, however, because of their great variation are of little value commercially. Their breeding value lies only in the range of variation which they provide in a selection program extending over a number of generations in which the aim is selection for the most desirable characteristics

of both breeds. Some breeders have capitalized on the advantages of outcrossing by involving three breeds, each of which contributes especially desirable characteristics. However, such crosses may be expected to involve a longer period of selection to arrive at the ultimate objectives.

Inbreeding in contrast to outbreeding, is the mating of closely related individuals. The closest form is brother-sister or parent-offspring mating. Carried on for 20 generations or more it leads to genetic uniformity. Opinion varies with regard to its use. In general, it is in bad repute because it usually is initiated with a stock previously outbred for many generations; such a stock is likely to carry a large pool of undesirable recessive genes covered up by the process of outbreeding. Some of these genes may be lethal, thus reducing viability and reproductive fitness as they are brought together by successive generations of inbreeding. However, as these recessive genes are observed and ruthlessly discarded, the strain in each generation tends to become more and more uniform. Inbreeding in itself does not create harmful genes; it only exposes those that are already present. At the same time, careful selection fixes favorable and desirable dominant and recessive characters so that uniformity is progressively increased. In the smaller laboratory animals which reproduce more rapidly than rabbits, a number of successfully isogenic (as alike as identical twins in man) strains have been successfully established for research purposes. This means lines which are so much alike that skin and other tissues can be successfully grafted or transplanted among them. The procedure followed is simple and straightforward if no complications arise. It consists of strict brother-sister matings usually done *without selection* (since successful reproduction and uniformity are the major objectives) for at least 20 generations. This leads to the complete set of many genes of the breed being alike, pair for pair, except for that part which determines sex. It will continue to be so as long as such breeding is continued except for possible occasional mutations which under normal environmental situations rarely occur. New genes must never be introduced or the 20-generation procedure must be repeated and then the chances of having the identical gene pool are very small.

The major difficulty with inbreeding, and this the breeder must weigh carefully before undertaking such a program, is that during the first 10 to 12 generations sterility, mortality, and undesirable abnormal variations are certain to be high, rendering the undertaking economically costly and even vulnerable to complete loss. Unless sufficient offspring are produced in each generation to insure that only the absolute best are retained (that is, some selection is exercised), the program may be hazardous. Where selection is practiced, inbreeding progresses more slowly from generation to generation, but more safely. Once deleterious genes are fixed in an inbred generation the damage can be repaired only by some form of outbreeding. Once the program is initiated, new genes cannot be added in any generation without undoing all uniformity previously achieved. It is therefore most important that the initial stock be of the highest quality, that is, contain the maximum number of favorable genes. Should there be any question as to this quality of foundation stock or its ability to produce, line breeding

(the mating of animals of less close relationship) may be desirable for a few generations. This will acquaint the breeder with his unknown recessive gene pool and at the same time, by selection, provide some form of concentration of the best genes. It can be done most rapidly by keeping the relationship to some one desirable ancestor high. Because a prolific male can affect many more offspring than a female in a given time period, several generations of backcrosses to any exceptionally vigorous and prolific male may do much to strengthen the initial gene pool before full brother-sister mating is begun.

It should be pointed out that a number of attempts have been made to inbreed the rabbit in this country and abroad over the past 25 years but thus far no completely isogenic strain exists. Achievement of inbred lines is the only means of securing genetic uniformity and, although it is a hazardous undertaking, the breeder who has obtained some degree of success by any system of close breeding will find inbreeding a challenging approach to further breed improvement. With the increasing usefulness of the rabbit in medical and biological research, the demand for truly isogenic strains is almost certain to become greater and greater; and breeders who do undertake production of such might find it not only a profitable investment, but would render a most valuable service to medical and biological science. In such an undertaking the value and importance of some training in genetics, nutrition, animal husbandry, and health cannot be overemphasized. Young people who are interested in such efforts will do well to seek such training as early in their education as possible.

THE RABBITRY AND ITS EQUIPMENT

Select rabbitry equipment that is adapted to your local conditions and to your proposed operations by reviewing the literature on the subject. *If possible, visit rabbitries and discuss problems with successful breeders.* Have your equipment ready when the first rabbits arrive.

Buildings

The type of building you need for housing the hutches will be determined by local building regulations, climatic conditions, and the amount of money you can invest. In planning your building and its equipment, emphasize comfort of the rabbits and convenience of the caretaker. The building should have a simple design, protect the rabbits from winds, rain, and bright sun, and provide light and fresh air (figs. 2, 3). Where mild climates prevail, hutches may be placed in the open but should have individual roofs and protection from the weather.

Sunlight helps maintain a sanitary condition in the rabbitry but whether it actually helps the rabbits themselves has not been determined. Rabbits apparently enjoy being in the sun where temperatures are low or moderate but it is not necessary that they receive direct sunlight. In fact, exposure to direct hot sun may have serious deleterious effects on rabbits.

In mild climates, hutches may be placed in the shade of trees or buildings or under a lath superstructure (fig. 2).

In hot climates, some cooling measures must be provided in addition to shade. This can be accomplished by the use of overhead sprinklers, or foggers placed within the building. Make sure that the building is adequately ventilated

and that the rabbits receive the benefit of prevailing breezes. In areas where strong winds and stormy weather prevail, you can put up hutches in a building that is open to the south and east; use curtains or panels to close up the building during inclement weather. Where you have extremely cold weather, more protection will be needed (figs. 4, 5).

Hutches

Provide individual hutches for mature rabbits. The hutches should be no more than 2½ feet deep so you can easily reach the rabbits, and 2 feet high. Make the hutches 3 feet long for small breeds, 3 or 4 feet for medium-size breeds, and 4 to 6 feet for giant breeds. All figures are for inside measurement. Whether you arrange the hutches in single, double, or triple tiers depends upon how much room is available. If you have enough room, waist-high, single-tier hutches are preferable as they are most convenient for observing the rabbits and will also save time and labor in feeding and management. The two- or three-tier hutches, necessary when space is limited, are not entirely satisfactory for caring for and observing the animals in the bottom and top tiers. The inconvenience of squatting or stooping to feed and care for rabbits in the bottom tier and of having to use a stool or ladder for the top row of a three-tier arrangement results in additional labor and time as compared to a single-tier arrangement.

Rabbits are more easily cared for in well-built hutches than in poorly constructed temporary ones. Self-cleaning, all-wire hutches (fig. 6) need no bedding and you can easily keep them in good condition.

BN 26084

N 45961

FIGURE 2.—Typical buildings used in areas of mild climate.

FIGURE 3.—Typical rabbitry in areas where the climate is hot.
(Courtesy of Small Stock Magazine.)

FIGURE 4.—Outdoor hutches used in Central States.

Metal Hutches.—Several designs of wire hutches are available commercially or you can build your own. Plans and specifications may be obtained from commercial firms who advertise in the various rabbit journals. A metal hutch that saves labor in caring for animals and is simply designed and economical to build is a combination two-compartment all-wire hutch.

An all-wire quonset-shaped hutch (fig. 6) has several advantages. It is easy to clean, neat in appearance, and requires less wire than a standard rectangular hutch.

The hutch features a door that opens up over the top. When open, the door does not occupy aisle space or interfere with feeding and cleaning operations. In addition, when this type of hutch is single-tiered at waist height, you can reach all the corners without placing your head and shoulders inside the door opening.

Quonset-shaped hutches can be adapted to fit any type of rabbitry where hutches are protected. They are most easily constructed in units—two hutches per unit.

BILL OF MATERIAL FOR TWO QUONSET-SHAPED HUTCHES IN ONE UNIT

The following material will be needed to build one unit containing two hutches—each hutch will be 3 feet long and 2½ feet wide:

Floor:
One piece of welded, 16-gage galvanized wire, 1- by ½-inch mesh, 3 feet wide by 6 feet long.

Top:
One piece of welded, 14-gage galvanized wire, 1- by 2-inch mesh, 4 feet wide by 6 feet long.

Ends and partition:
Three pieces of welded, 14-gage galvanized wire, 1- by 2-inch mesh, 1½ feet wide by 2½ feet long.

11527–D

FIGURE 5.—Semienclosed hutches for use in cold climates.

N45939

FIGURE 6.—Quonset-shaped, all-wire hutches with counterset nest box and hopper feeders.

Doors:

Two pieces of welded, 14-gage galvanized wire, 1- by 2-inch mesh, 1½ feet wide by 1 foot 8 inches long.

Miscellaneous:

Steel rod.—5/16-inch round steel rod, 8 feet 11 inches.

Two pieces, 2 feet 6½ inches long for nest supports.

One piece, 3 feet 10 inches long, for reinforcing the front of the hutch.

Wire.—No. 12 galvanized, 21 feet 7½ inches.

Three pieces, 4 feet long, for edging around ends and partition.

Two pieces, 9¼ inches long, for vertical nest support.

Two pieces, 1 foot 4¼ inches long, for horizontal nest support.

Two pieces, 2 feet long, for feeder yokes.

Wire.—No. 9 galvanized, 7 feet 4 inches.

Two pieces, 1 foot 8 inches long, for additional support at the ends of the hutch.

Two pieces, 2 feet long, for reinforcing the door openings.

Fasteners:

100 hen-cage clips, small size, for fastening the floor, top ends, and partition.

25 hen-cage clips, large size, for door hinges and for fastening the No. 9 wire.

30 hog rings, No. 101, for fastening the 5/16 inch steel rod to the floor.

Door latches.—Any standard latch, or fastener, may be used.

In constructing a unit of two

hutches, it is recommended that the floor be laid out first, 3 inches to be bent up on either side (the front and rear of the hutches), and openings cut for the counterset nest boxes. The 3 inches can be bent up on the sides with a metal brake or improvised homemade tooling. In cutting openings for the nest boxes be sure to leave approximately 1½ inches of flooring at the front of the hutch for suspension of the nest boxes.

The partition and ends should then be shaped from 1-inch by 2-inch wire by using a template. Allow ⅝-inch protrusion beyond the edge of the template and bend these wires around the No. 12 edging wire. At this time, some No. 12 wire may be fastened to the bottom of the ends and partitions for reinforcement. The ends and partitions can then be laid in position on the floor, and fastened to the floor with hen-cage clips.

Next make the top from 1-inch by 2-inch wire, cutting openings for the doors and feeders. Lay the top over the floor, ends, and partitions, and fasten at the rear with hen-cage clips spaced approximately every 5 inches. Raise the front edge of the top until it is even with the 3-inch raised front edge of the flooring and fasten with hen-cage clips. Now, reach in one end and raise one end enclosure into position, fastening it to the top with hen cage clips. Repeat this process with the center partition and other end enclosure. This will automatically form the quonset-shape top over the ends and partition. The raised front edge can then be cut for installation of the feeders, the doors and nest boxes can be installed, and the hutch is ready for use.

When two or more units (four or more hutches) are built and placed end to end, a saving of one end enclosure can be made for each

unit built by using the following procedure:

1. Cut the wire mesh for the floor and top of the first unit 6 feet, 1 inch long.

2. On the first unit, fasten the partition 3 feet from the left-end enclosure—fasten the right-end enclosure 3 feet to the right of the partition. This will leave a 1-inch overhang to connect to the next unit.

3. On all additional units, cut the floor and top wire 6 feet long and fasten the partition 35 inches from the left end; fasten the right-end enclosure 3 feet to the right of the partition, leaving 1 inch of floor and top extending beyond the right-end enclosure. These end enclosures become partitions when units are added.

4. Use hen-cage clips to fasten the units into one continuous line.

The hutches can be installed in several ways. Suspension from the rafters or ceiling of a shed is the most practical method because it eliminates all supports beneath the hutches. Heavy wire or light lumber can be used to hang the hutches. If a dewdrop water system is used in the rabbitry, the hutches can be fastened to the water pipe for rear support.

If they are not placed within a shed, the hutches can be supported by a frame on legs. However, the hutches will require some type of cover to protect the rabbits from rain, sun, and wind.

Wooden-Frame Wire Hutches.— Though not so durable as the all-wire hutch, the wooden hutch with woven-wire sides and ends permits good circulation of air. It is more sanitary than a solid hutch.

Hutches may be supported in several ways. If you use corner posts, make them long enough so that you can clean underneath and do other work around the hutch. You can support a hutch by resting it on a crosspiece nailed between the studs that support the shed, or you can hang it from the rafters or ceiling of the shed with heavy wire or light lumber.

Semienclosed Hutches.—The semienclosed hutch is constructed with ends and back of wood (figs. 4, 5). An extended roof gives added protection. You can use this hutch in outdoor rabbitries in cold climates.

Another satisfactory type of hutch, which is light, movable, and inexpensive, is shown in figure 7.

Rabbits kept in hutches made of wooden frames and wire need additional protection in cold climates.

Hutch Floors.—Several types of floors are used in hutches, and each has its particular merit.

Wire mesh floors are used extensively where a self-cleaning type is desired. They are a necessity in commercial herds, where it would be impossible to provide enough labor to keep solid floors in a sanitary condition. In installing this type of floor, examine the wire for sharp points which result sometimes

from the galvanizing process. Always put the smooth surface on top. Solid floors should slope slightly from the front of the hutch to the rear to provide proper drainage. You can use hardwood slats, 1-inch wide and spaced ⅝-inch or ¾-inch apart. A combination of solid floor at the front part of the hutch and a strip of mesh wire or slats at the back may be used.

Feeding Equipment

It is desirable to use feed crocks, troughs, hoppers, and hay mangers that are large enough to hold several feedings, to save time in filling. Use a type that will prevent waste and contamination of the feed.

Crocks.—Crocks especially designed for rabbit feeding, which are not easily tipped over, have a lip that prevents the animals from scratching out and wasting their feed. The chief objection to these is that the young rabbits get into them and contaminate the feed.

83091-B'

FIGURE 7.—An economical hutch of light construction, which can be moved from place to place.

Hay Mangers and Troughs.— Hay mangers with troughs to prevent wastage may be incorporated into hutches, where hays or green feeds form a part of the diet. The troughs also can be used for supplemental grains or home-grown feeds. The troughs may be constructed so that they can be pulled out of the hutch for cleaning, filling, and disinfecting. Guards placed on the feed troughs and spaced just far enough apart to allow mature animals to feed, will help keep young rabbits out of the troughs and from contaminating the feed.

Hoppers.—Feed hoppers of the proper design and size save considerable time and labor. These can be constructed from metal, wood, masonite, or other readily available materials. They should hold at least several days' supply of feed and be placed within the hutch or suspended on the outside (fig. 6). The opening through which the rabbits obtain feed should be not more than 4 inches above the hutch floor so that young rabbits can readily obtain feed. An inexpensive feed hopper that will hold about 15 pounds of pellets or grain can be made from a common square 5-gallon can (figs. 8, 9). First, cut off the top. Then cut holes in two opposite sides. If the hopper is to be hung on the side of the hutch, cut a hole on one side only. The holes should be 4 inches high, 4 inches from the bottom, and 1 inch from each side. Bend the rough edges inward to give a smooth edge all around and to add rigidity. Take a 1- by 4- by 13½-inch board and cut it diagonally into two equal triangular pieces. Use these as supports to the baffle boards, which are nailed to them.

The baffle boards, of ½-inch plywood, should extend 1 inch below the bottom of the side openings of the can. The space between the lower ends of the baffle boards permits the grain or pellets to flow

N45944

FIGURE 8.—Feed hopper (self-feeder) constructed from 5-gallon can.

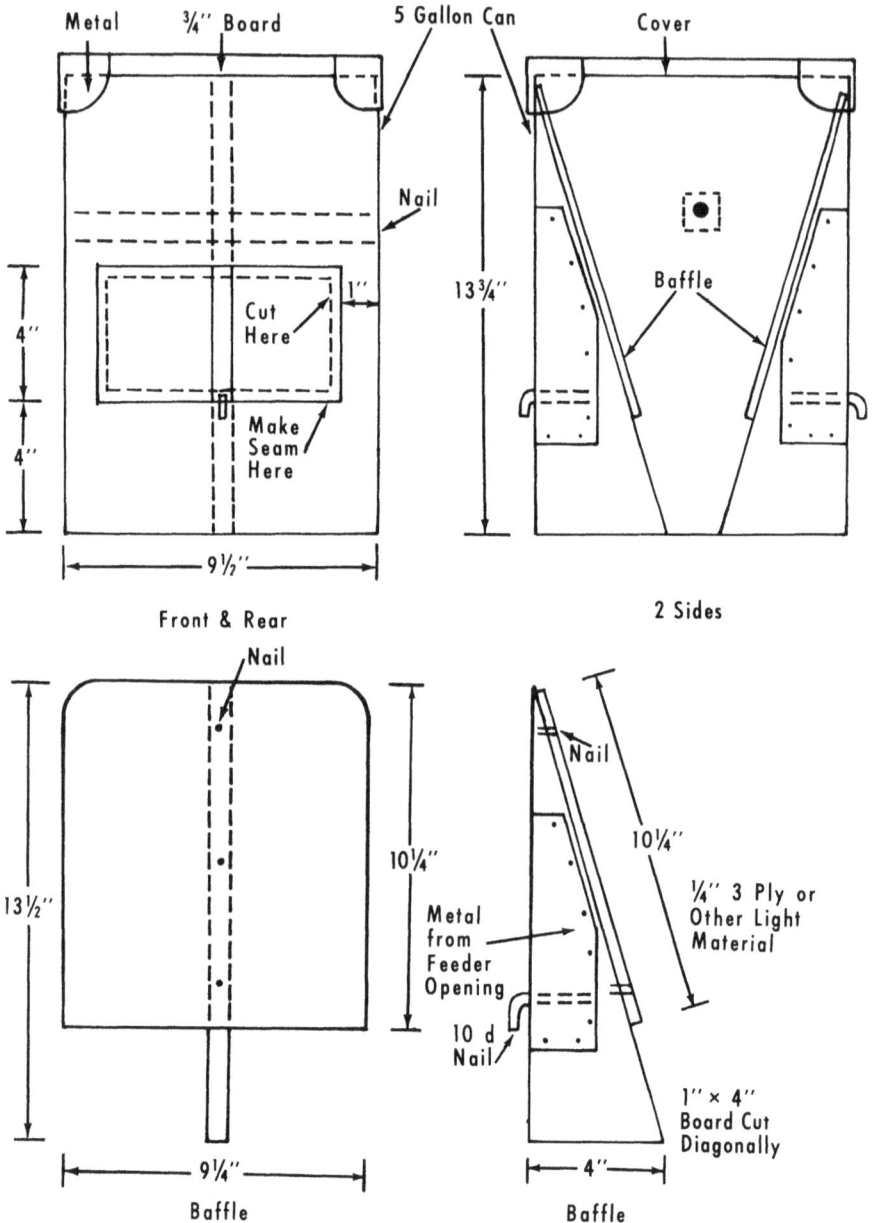

FIGURE 9.—Details of feed hopper made from 5-gallon can.

down as the rabbits eat. Make the baffle boards to fit snugly against the sides of the can so feed cannot slip by. Mount the top corners of the baffles so that each baffle will rest against the top edge of the can.

Cover the exposed edges of boards with tin to prevent gnawing. Put a finishing nail in the outer edge of the triangular piece supporting the baffle, and bend the nail to hook over the lower lip of the opening to hold it and the baffle in place.

You can save hutch floor space by using a hopper with a feed opening on one side only and by placing the hopper only part way into the hutch. Cut an opening large enough to accommodate the hopper in the side of the hutch. Then wire the top of the hopper to the hutch for support. One short baffle on the side opposite the hopper opening will keep feed out of the rear corners.

A one-compartment feed hopper is used when only one kind of feed is given. When mixed feed that the rabbits can separate is offered in the hopper, the feed will be selectively consumed. The rabbits scratch out and waste the part they prefer not to eat. You can prevent this waste by using a hopper with individual compartments for each feed.

Equipment for Watering

Rabbits should have clean, fresh water at all times.

Crocks.—Half-gallon water crocks are still used rather extensively. Fasten them in the hutches so that the rabbits will not tip them over. If a part of the crock extends through the front wall of the hutch, you can refill it without opening the hutch door. Clean and disinfect the crocks periodically.

Coffee Cans.—Coffee cans are especially useful for watering rabbits during cold weather because you can easily break and remove the ice. Cans are, however, easily tipped over unless you fasten them to a board.

Automatic Watering System.— Automatic watering systems are widely used in commercial rabbitries (fig. 10). They are better than

water crocks or coffee cans. They eliminate the tedious and time-consuming chores of washing, disinfecting, rinsing, and filling. They supply fresh, clean water for the rabbits at all times. When an automatic watering system is properly installed, dirt and fur will not collect in it and plug it up. In cold climates, an automatic watering system must be protected against winter freezing unless the hutches are in a heated enclosure. Protection may be obtained through the use of heating cables wrapped around, or running through the water pipe. If winter temperatures are not too severe, protection against freezing can be obtained by having valves at the ends of the water lines and allowing water to dribble throught the pipes during short periods of subfreezing temperatures.

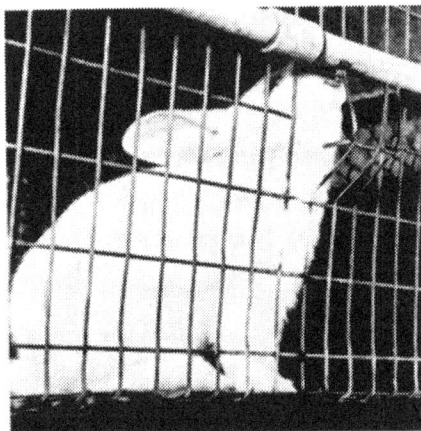

12767A

FIGURE 10.—Young rabbit drinking from an automatic waterer.

If you can cut and thread pipe, you can install an automatic watering system. Conventional systems sold by rabbit and poultry supply houses consist of a pressure-reducing tank equipped with a float valve, a $\frac{1}{2}$-inch supply pipe, a watering unit for each hutch, and valves. The valves are used to bleed out air bubbles, to drain the system as needed, or to shut off the water. If the water contains sediment, a half-barrel can be advantageously used instead of the standard pressure-reducing tank. The outlet for the supply pipe can be installed several inches above the bottom of the barrel. The sediment will then collect below the outlet pipe and will not get into the system and clog it. Other sediment traps, installed between the tank and the supply pipe to the hutches, can be used with any type of tank.

One-gallon tanks or smaller tanks sometimes are used where the weather is warm. Such tanks are emptied more often. The constant flow of water in and out of the tank keeps fresh cool water before the rabbits at all times.

Install the pressure tank 1 foot or more above the highest hutch. If the supply pipe is raised to clear the feeding alleys, then install the tank about 1 foot above this highest point.

Raised supply pipes may require vent pipes to keep air bubbles out of the system. Install the vent pipe at the highest point in the supply line. See that the open end is at least 1 foot above the water level in the tank. If it is necessary to change the level of the supply line from one row of hutches to another, use a piece of rubber hose to make the connection.

Determine the correct height for the tank by fastening a rubber hose to the tank outlet and then to the supply pipe. Raise or lower the tank until the valves, or dewdrops, from which the rabbits drink have the proper tension. If there is too much tension or pressure on the valves, the rabbits will not be able to trip them. Under too little tension, the valves will drip.

The proper height for the water valve is 9 inches from the hutch floor for medium and heavy breeds and 7 inches for the smaller breeds. The pipe may be hung on the outside and at the back of the hutch so no water drips on the rabbits and the hutch floor. An opening in the back of the hutch will permit the rabbit to use the valve (fig. 10).

When hutches are back to back use one pipe for supplying water to both hutches. Use a four-way outlet and short nipples for installing the valves.

You can install one drinking valve for each hutch by drilling and tapping the supply pipe and screwing the valve into it.

If you are not equipped to make the plumbing installation, substitute a $3/4$-inch rubber hose for the $1/2$-inch supply pipe. Cut a hole in the hose and screw in the valve. Plastic pipe may be used in a similar manner. If a rubber hose or plastic pipe is used it should be hung on the outside of the hutch to minimize possibilities of damage due to chewing or gnawing.

Check the automatic watering system periodically, especially when you put a rabbit in a hutch that has been unoccupied for several days. When valves are not used—even for a few days—minerals in the water may cause them to stick.

Rabbits learn to use the system readily, even young just out of the nest box.

Nest Boxes

No one type of nest box is best suited for all conditions, but all should provide seclusion for the doe at kindling and comfort and protection for the young. Nest boxes should be large enough to prevent crowding and small enough to keep the young together. All types should provide good drainage and proper ventilation. Two general kinds have been used extensively—the box type and the nail-keg type. If a nail keg is used, nail a 1- by 6-inch board across the open end of the keg, so that it covers one-third to one-half of the opening. To keep the keg from rolling, extend the board a few inches beyond the sides of the opening. Drill several 1-inch holes in the closed end of the keg for ventilation, and some ¼-inch holes in the bottom for drainage.

Since nail kegs have become difficult to obtain, apple and pear boxes are frequently used. These may be fitted with tops or left open. In either event, an opening should be cut in one end at the top, or a portion of one end removed, to provide easy access for the doe and young. As an alternative, one end may be fitted with removable boards, or slats, so that as the young begin leaving the nest, panels may be removed to allow the young to reenter the nest box. Metal nest boxes also are available but have the disadvantage, in some climates, of being cold or collecting condensation of water vapor.

Another type of nest box increasing in popularity is the counterset type, where the box is recessed below the hutch floor (figs. 6, 11). These may be placed at the front of the cage and fitted with drawers for access from the exterior of the hutch. They have the advantages of providing a more natural environment, since rabbits are burrowing animals, and of allowing the young easier access if they should be displaced from the nest at an early age. The young can jump out of the standard nail-keg or apple-box nest, but they often cannot jump or climb back in. This means that some of the young may go hungry when the litter becomes divided. The doe usually nurses her young at night or in the early evening and morning hours. If the litter is divided, the doe will either nurse the young in the nest or those on the hutch floor. She will not nurse

N 45948

FIGURE 11.—Counterset nest box and drawer as illustrated in figure 6.

both groups, nor will she pick up the young and return them to the nest.

The counterset nest boxes are easier to keep clean than the apple-box and nailkeg nest boxes because the inner drawers of the counterset nest box can be slipped out for washing and disinfecting. These drawers also can be interchanged from one hutch to another. When the young no longer need the inner drawer, it can be left out to provide more space in the hutch.

BILL OF MATERIAL FOR NEST BOX AND DRAWER

Nest box:

Sides.—Two pieces of lumber,[2] 1 by 12 by 17 inches.
End.—One piece of lumber, 1 by 8 by 12¼ inches.
Door.—One piece of lumber, 1 by 12¼ inches.
Winter enclosure.—One piece of lumber, 1 by 8 by 12¼ inches.
Cover.—One piece of ⅛-inch hardboard, 12 by 12¼ inches.
Bottom.—One piece of 16-gage galvanized wire, 1- by ½-inch mesh, 12 by 18 inches.

Nest drawer:

Sides.—Two pieces of ⅛-inch hardboard (tempered), 7¼ by 16¾ inches.
Ends.—Two pieces of lumber, 1 by 8 by 10 inches.
Bottom.—One piece of ⅛-inch hardboard (tempered), 8½ by 16¾ inches.

Miscellaneous:

Nails.—Use sixpenny or eightpenny nails to fasten the end, top, and sides of the nest box, eightpenny to fasten the nest drawer, and 1¼-inch roofing nails (large head) to fasten

_____
[2] No. 2 construction knotty pine or 1-inch box lumber.

the wire bottom to the nest box.
Protecting strips.—To prevent chewing and splintering, nail 30-gage galvanized sheet metal, bent to form a ½- by ⅝-inch angle, to the exposed edges of the nest box and drawer.

Hinges.—Two 1-inch strap hinges for the door.

In shaping the sides of the nest box for the slanted roof, you can use the piece of lumber cut from the rear of each side to build up the front. The completed sides should be 17 inches long, and should slant from 16 inches tall in the front to 8 inches tall at the rear (fig. 6).

Suspend the completed nest box in the hutch by the cradle of No. 12 wire at the rear and the three remaining strands of hutch flooring in the front. The cradle of No. 12 wire can be made in three sections to fit down each side of the box and under the bottom, or in one long piece. In either case it is merely hooked onto the hutch flooring next to the nest box on one side, passed down and across beneath the box and up the other side to again hook on the hutch floor. This provides adequate support for the rear of the nest box. Slip the three strands of flooring into notches cut into the front end of the nest box just above the door.

To prevent the nest box from slipping to the rear so that the floor wire at the front end no longer acts as a support, the side boards of the nest box can be cut so as to extend a little above the back board of the nest box. Then as the back board comes up under the hutch floor, these side boards project a little above the floor and prevent the nest from being pushed to the rear.

To help keep the nest dry, cut some ¼-inch drain holes on the bottom of the nest-box drawer.

Losses of young rabbits kindled in winter can be largely prevented if you furnish proper nesting accommodations. If a doe reacts normally to her newborn litter by pulling enough wool to make a warm nest and feeding her young, and if the nest box is well insulated, the young can survive temperatures as low as 15° to 20° below zero.

You can make a good type of winter nest box by placing a standard size nest box inside a larger box. Pack straw into the space of 3 inches or so on all sides except entrance and top. A lid of ordinary box wood covered on the under side with two thicknesses of paper will supply the necessary top insulation. Make two or three holes, ½- to ¾-inch in diameter, in the lid at the end opposite the opening to the nest box, for ventilation and to prevent condensation within the nest box. On the bottom of the inner box, put one or two layers of corrugated cardboard or several thicknesses of paper to keep the newborn litter from coming in contact with the cold boards. Fill the nest box so completely with new, clean straw that the doe will have to burrow into it to form a cavity for a nest. Inspect the box daily for the first 3 or 4 days. If the cardboard or paper becomes damp from accumulated moisture, remove it promptly. Replace it if cold weather continues. A simpler nest box for use in winter consists of a single box lined completely with one or two layers of corrugated cardboard and filled with straw.

FEEDS AND FEEDING

Success in raising rabbits is impossible if you do not give enough attention to diets and provide wholesome feeds in adequate quantity each day.

Feed is one of the biggest items of expense in raising rabbits and *each herd presents an individual* problem. Select diets that are suited to the needs of your rabbits, whether you buy commercially prepared mixtures or pellets, or mix feeds yourself.

Feed Requirements

Rations for dry does, herd bucks, and developing young should provide the following:

	Percent of ration
Crude protein	12 to 15
Fat	2 to 3.5
Fiber	20 to 27
Nitrogen-free extract	43 to 47
Ash or mineral	5 to 6.5

Rations for pregnant does and does with litters should contain more protein. Their rations should include:

	Percent of ration
Crude protein	16 to 20
Fat	3 to 5.5
Fiber	15 to 20
Nitrogen-free extract	44 to 50
Ash or mineral	4.5 to 6.5

The nutrient contents of common rabbit feeds are shown in table 2, and daily requirements for various weights of rabbits are shown in table 3. Further information on nutrient contents may be obtained from Morrison's Feeds and Feeding (10)[3] and National Research Council Publication No. 1194 (12).

The protein content of rations is important in development of

_____
[3] Italic numbers in parentheses refer to the Literature Cited, p. 69.

TABLE 2.—*Digestible protein and total digestible nutrients of some common rabbit feeds*

[Dry roughages and concentrates on an air-dry basis]

Feed	Dry matter	Digestible crude protein	Total digestible nutrient
DRY ROUGHAGES	*Percent of ration*	*Percent of ration*	*Percent of ration*
Alfalfa hay, common	90	11	40
Alfalfa hay, very leafy	90	16	58
Bluegrass hay	92	8	31
Clover hay, red	88	7	43
Lespedeza hay, annual	89	8	39
Sorghum fodder, milo	88	6	35
Oat hay	88	5	26
Peanut hay, without nuts	91	6	46
Soybean hay	88	10	43
Sudangrass hay	89	6	43
Timothy hay	89	3	32
Vetch hay (common)	89	10	46
GREEN ROUGHAGES, ROOTS, AND TUBERS			
Alfalfa	21	4	15
Cabbage, aerial portion	9	2	9
Carrots, roots	12	1	10
Clover	20	2	13
Rutabagas, roots	11	1	10
Sweetpotatoes, roots	32	1	28
Turnips, roots	9	1	8
CONCENTRATES			
Barley grain	89	10	70
Beet pulp, dried	90	4	70
Bread, dried	64	8	65
Brewers' grains, dried	93	23	58
Buckwheat grain	88	7	70
Corn, grain dent #2	85	7	82
Cottonseed meal	92	32	66
Linseed meal	91	31	70
Milk, cows	13	3	16
Milk, dried	96	26	117
Oats, grain	90	9	65
Peanut meal	93	39	85
Sorghum grain, milo	89	8	84
Soybean meal	91	40	82
Soybean seed	90	33	98
Wheat grain	89	11	79
Wheat bran	90	14	57

young, for maintaining the breeding herd, and for wool production. It also is a factor in the quantity of food required for a certain gain in live weight. Adding the proper quantity of protein supplement to a ration composed of grains and hay increases the rate of growth of young rabbits 13 to 20 percent and effects a saving of 20 to 25 percent in the quantity of feed required for a unit of gain.

Protein is the most expensive part of the feed, but the propor-

TABLE 3.—*Daily nutrient requirements of rabbits per animal*

[All feeds or rations are based on air-dry weights]

Phase of production and body weight	Total feed	Total digestible nutrients	Total protein	Digestible protein
	Lb.	Lb.	Lb.	Lb.
Normal growth, does or bucks, 4 to 9 lb , average 6.5 lb	0 32	0. 19	0. 05	0. 03
Normal growth and fattening, does or bucks:				
4 lb	. 25	. 16	. 04	. 03
5 lb	. 30	. 19	. 05	04
6 lb	. 34	. 22	. 05	. 04
7 lb	. 38	. 25	. 06	. 05
Maintenance, does or bucks.				
5 lb	. 20	. 11	03	02
10 lb	. 33	. 18	04	. 03
15 lb	. 45	. 25	. 05	. 01
Pregnant does:				
5 lb	. 25	15	. 04	03
10 lb	41	24	. 06	. 05
15 lb	. 56	. 33	. 08	06

tions recommended are those that have proved most economical. The upper limits suggested give better results than the lower. *There is no danger in feeding higher levels of protein than recommended provided the ration is adequate in all other ingredients.* Thus, if your herd is small or if it would be difficult to feed two rations, you can give feed intended for pregnant does and for does with suckling litters to the entire herd.

Many rabbit raisers will have homegrown grains and hay or will be able to purchase them locally. These feeds in their natural form are satisfactory if you use additional protein to balance them properly. Feed them in separate compartments of a self-feeder or use the plant - protein supplements—soybean, peanut, sesame, cottonseed, and linseed meals in the pea-size cake, flake, or pelleted form—with whole grain to make up the concentrate part of the ration. If you hand-feed the mixture, use a container that prevents the rabbits from scratching out

and wasting the feed. If you use finely ground mill products in the mixture, dampen the feed just before feeding to prevent the fine meals from settling out and being wasted.

Hay

For your rabbits, choose hay that is fine stemed, leafy, green, well cured, and free from mildew or mold.

If you feed whole, coarse hay, a good deal will be wasted. The rabbits will pull a stem out of the hay manger, eat part of it, and drop the rest. To prevent some of this waste and to put the hay in a more convenient form for feeding, cut it into 3- or 4-inch lengths.

The legume hays, such as alfalfa, clover, lespedeza, cowpea, vetch, kudzu, and peanut are palatable and make good feed for rabbits. The carbonaceous hays, such as timothy and prairie, and hays made from johnsongrass, sudangrass, or dallisgrass, are less palatable than legume hays, but are valuable for feeding where legume hays are not readily available.

The grass hays ordinarily contain only about half as much protein as legume hays. If you feed them, include more protein supplement in the diet. If they are cut before the plants are in bloom, when the stems are fine and there is a high proportion of leaf, the grass hays are much more suitable for feeding. They have a higher protein content at this time but they never contain as much protein as legume hays.

Hays furnish bulk or fiber in addition to nutrients. Rabbits fed insufficient bulk have soft droppings that mash on the hutch floor and cause increased labor in keeping the hutch clean. They also may chew their fur. If you feed young rabbits too much bulk they will not get enough nutrients for rapid growth and market finish.

Green Feed and Root Crops

Rapid-growing plants, such as grasses, palatable weeds, cereal grains, and leafy garden vegetables free from insecticides, are high in vitamins, minerals, and proteins, and make excellent feeds, especially for the breeding herd. Use them in the diet when they fit into the management program.

Root crops, such as carrots, sweetpotatoes, turnips, mangels, beets, and Jerusalem-artichokes, are desirable feeds throughout the year, and are particularly good in winter when green feeds are not available.

Fresh green feeds and root crops should be used as supplements to the concentrate part of the diet. You will get best results when you use variety. Fresh feeds contain 90 percent or more of water. Use them only as supplements to grain or pellets when choice carcasses are

desired. You can use them to maintain mature animals that are not in production.

Feed root crops and green feed sparingly to rabbits that are unaccustomed to them. There is no danger in feeding fresh green feed that is wet with dew or rain. Do not use feed that has been piled and become heated.

Place green feed in a hay manger; never throw it on the floor of the hutch. Contaminated feed may cause digestive disturbances or re-infect rabbits with internal parasites. Remove any feed that is not readily consumed.

Grains and Milled Feeds

Use oats, wheat, barley, the grain sorghums, buckwheat, and rye as whole grains or as milled products. You can feed the softer varieties of corn whole, but there will be considerable waste of the flinty varieties unless you feed them in meal or cracked form. The grains are quite similar in their food values and you can substitute one for another on a pound-for-pound basis without materially altering the nutritive value of the ration.

Milled-wheat products such as bran, middlings, shorts, and red-dog flour, and byproducts from manufacturing foods from other grains for human use may be included in mash mixtures and pellets.

Rabbits eat sunflower seeds readily, but because they have a much higher value for other uses they seldom are included in rabbit diets.

Protein Supplements

Soybean, peanut, sesame, cottonseed, and linseed meals are rich in protein and desirable for balancing rabbit rations. These feeds in meal form are used in mashes and pelleted rations but are unsatisfactory for mixing with grains. They will settle out of the grain mixture and be largely wasted. The pea-size cake, the flake form, or the meals made into a pellet are satisfactory for use with whole grains. If their protein content is the same, the meals in pea-size cake, flake, or pelleted form provide approximately the same nutritive value. Make your selection on availability and cost. Use fresh plant-protein supplements.

Although soybean seeds contain approximately 36 percent protein and 18 percent fat, the meal from the seeds, with fat extracted, has as much as 45 percent protein and 1 to 5 percent fat. If there is an oil mill nearby, you may be able to exchange homegrown soybeans for the meal. Rabbits do not eat the seeds readily—feed only about 1 pound of them for each 10 pounds of grain. Using this proportion of soybeans in a whole-grain legume hay diet will improve the protein content slightly but not enough for maximum growth.

Some caution should be observed in using cottonseed meal as a protein supplement. Untreated cottonseed meal contains gossypol, a substance which is toxic to rabbits. Therefore, only degossypolized meal should be used. Recent evidence from the U.S. Rabbit Experiment Station indicates that degossypolized cottonseed meal is a suitable replacement for soybean meal at levels up to 7 percent of the diet.

Miscellaneous Feeds

Dry bread or other table and kitchen waste (except meat and greasy or sour foods) are acceptable to most rabbits. When used as supplements to grain and roughage or pelleted rations, they add variety to the diet. When the cost is not prohibitive, cow's or goat's milk may be used in the diet. If the milk is not sour or contaminated, it will not cause digestive troubles. Dry bread mixed with milk is a satisfactory feed for does with young litters and for rabbits being conditioned for shows.

Pelleted Rations

Many brands of pelleted rations are on the market. Ingredients and proportions vary but they are usually made according to recommended specifications of nutrient or feed content. Follow the advice of the manufacturer.

Pelleted rations require little storage space and are easily fed. In some localities they are the only rabbit feeds available.

There are two types of pelleted diets—the all-grain pellet to be fed with hay and the complete pellet (green pellet). *The complete pellet usually contains all the food elements necessary for a balanced diet.*

The choice between a home-mixed feed or a pelleted feed will depend on the availability and relative cost, and how much time you have for preparing the ration and feeding the herd.

Pellets should be $1/8$- to $3/16$-inch in diameter and $1/8$- to $1/4$-inch long. If pellets are too large, small rabbits cannot get them in their mouths. The rabbits bite off a part of the pellet and drop the rest. The discarded part is lost through the wire hutch floor or is left to become contaminated on solid floors.

It is usually impractical for you to pellet your own rations.

Salt

Salt is necessary in the diet. Put small blocks or salt spools in the hutch so the animals can feed at will, or add 0.5 to 1.0 percent salt to mixed feed or pellets. Salt blocks or spools will cause corrosion of any metal with which they are in contact, so it is not advisable to use them in all-wire hutches. In areas where the soil is deficient in certain mineral elements, use mineralized salts, as fed to other farm animals, in rabbit rations.

Water

Rabbits need ready access to fresh, pure water at all times. In summer, they require large quantities. A 10- to 12-pound doe and her 8-week old litter of seven will drink about a gallon of water in 24 hours.

Preparing and Storing Feeds

Whole grains are satisfactory for feeding rabbits. Milled products, whether rolled, cracked, or ground, lose some of their food value and apparently become less palatable if stored for any length of time, especially during the summer. Coarse hay is more convenient to feed and less wasteful if you cut it into 3- or 4-inch lengths. Cutting the hay you feed to Angora rabbits helps keep the wool clean.

Sometimes you can save money by storing home-grown feed or feed purchased as it is harvested. Store it in rodent- and insect-proof containers.

Protect grains, pellets, hay, or other feeds and bedding materials from contamination by cats or dogs; otherwise the rabbits may become infested with a cat or dog tapeworm.

Methods of Feeding

Two methods of feeding are in general practice. One entails placing a measured amount of feed in feed crocks or troughs each day, and is referred to as "hand feeding." The other utilizes a hopper, or self-feeder, which holds several days' supply of feed, and is referred to as "self-feeding" or "full-feeding," since feed is available to the rabbits at all times and they can feed at will. Crocks or troughs may be used for full-feeding, but will have to be filled more frequently than hoppers and are more susceptible to waste and contamination. The hopper feeding system saves time and labor and prevents waste and contamination if the hopper is properly constructed.

Full-feeding in crocks or troughs produces about the same results as hopper feeding, provided you feed the animals all they will consume each day without waste. If you cannot give close attention, hopper feeding will give better results.

Whether, in hand feeding, a herd of rabbits should be fed 1, 2, or 3 times a day is largely a matter of personal preference and convenience. Regularity is more important than the number of feedings. Rabbits eat more at night than during the day, especially in warm weather.

Full feeding insures rapid growth and economical development of young to weaning. Full-fed rabbits generally require less feed than hand-fed rabbits to produce each pound of live weight because they eat frequently and slowly and chew their food thoroughly.

Occasionally a rabbit goes "off feed." When this happens, reduce the quantity of the ration. The offer of a tempting morsel of carrot, bread and milk, or fresh green feed may induce the rabbit to begin eating again.

Feeding Dry Does, Herd Bucks, and Juniors

You can maintain mature dry does and herd bucks not in service on hay alone if you freely feed a fine stemmed, leafy, green-colored, legume hay. If you feed coarse legume hays or carbonaceous hays, feed each 8-pound animal 2 ounces (⅓ cup) of a grain-protein mixture or an all-grain pellet several times each week. For rabbits of other weights, adjust the quantity. For example, feed 3 ounces to a 12-pound animal.

Feed herd bucks in service the same quantity of concentrates and give them free access to choice hay; or provide them with 4 to 6 ounces of a complete pellet daily. Regulate the amount to keep them in good condition and to assure that they do not become too fat.

For developing junior does and bucks, regulate the concentrate portion of the diet so that the animals will grow and be in good condition when they are ready for breeding. With the medium-weight breeds (9 to 12 pounds at maturity), hold the grain-protein mixture or the all-grain pellet on a daily level of 2 to 4 ounces and allow free access to a good-quality hay. As the rabbits develop, they will eat more hay to get the additional nutrients required for growth. When using a complete pellet, 4 to 6 ounces daily should be sufficient. Take precautions to prevent juniors from becoming too fat when they are fed concentrates or a complete pellet.

An alfalfa pellet, consisting of 99 percent No. 2 leafy, or better grade, alfalfa meal (15 to 16 percent protein) and 1 percent salt, may be full fed to developing junior does and bucks as the only feed from weaning until they are to be placed in the breeding herd. In the event that alfalfa pellets are unavailable through local mills, a coarse crumble or turkey-grind crumble, composed entirely of alfalfa, may serve as a satisfactory feed for developing stock. If a crumble is used it might be advisable to place a small salt block or spool in the hutch, though there is evidence from trials at the U.S. Rabbit Experiment Station that the animals may do without the extra salt for the few months before they are placed in the breeding herd.

Note: Nutritive value of diets, and daily feed requirements of individual rabbits, vary. Observe the condition of your individual rabbits and increase or decrease quantities of feed to obtain desired physical condition.

Feeding Pregnant and Nursing Does

To feed a doe properly, it is necessary to know definitely whether she has conceived. Palpating (feeling for the developing young in the uteri) at 12 to 14 days following breeding is a quick and accurate method of determining pregnancy (see p. 35).

After mating, you may maintain junior and mature does in breeding condition on good-quality hay or hay pellets until you have determined that they are pregnant. If your herd is receiving only complete pellets, restrict the amount that bred does receive daily to that quantity which will keep them in the desired physical condition until pregnancy is determined. Full-feeding complete pellets to a bred doe will cause her to put on too much flesh if she fails to conceive. If a doe fails to conceive as determined by palpation, breed her again and feed only hay, or restricted amounts of complete pellets, until she is pregnant. When she is diagnosed as pregnant, give her all the concentrates she will eat plus good-quality hay, or all the complete pellets she will eat for the remainder of the gestation period. You can provide the concentrates in the form of grain and a protein pellet or all-grain pellets. All-grain pellets have the necessary amount of plant-protein supplement incorporated with grain and salt to make a complete feed when fed with a good quality hay. *The general practice is to feed pelleted complete feed.*

Sudden changes in rations fed during the gestation period may cause some does to go "off feed." If they fail to eat necessary nutrients for too long a period of time, abortion or young that are dead at birth may result. Gradually change over a new ration by feeding one-fourth new ration and three-fourths old ration for 3 to 4 days, one-half new ration and one-half old ration for 3 to 4 days, and then three-fourths new ration and one-fourth old ration for 3 to 4 days.

After the doe kindles, she can be fed in the same manner as before, until the young are weaned when about 2 months old. From the day of kindling feed her all she readily will consume without waste, or a grain-protein mixture and hay, an all-grain pellet and hay, or a complete pelleted feed until the litter leaves the nest box. As the litter develops, feed the doe and litter greater quantities or full-feed them to insure maximum growth of the young. If you use a feed hopper and the hutch is small (less than 10 square feet of floor space for a 10- to 12-pound doe), placing a hopper in it with the nest box may make it too crowded. Full-feed the doe using a crock or trough until the nest box is removed, then introduce the hopper. Inspect the hopper occasionally to make sure that feed is always available.

COPROPHAGY

Rabbits re-ingest part of their food, usually in the early morning, when they are unobserved. They re-ingest only the soft matter that has passed through the digestive tract. Investigators have called this trait "pseudo-rumination," from the characteristic of ruminants (cows, sheep, and others) of chewing the cud, which is food regurgitated and chewed again. Most rabbit breeders are unaware of this practice. Some who have observed it believe it indicates a nutritional deficiency. It is, however, normal in rabbits and may actually enhance the nutritive value of the feed by virtue of a second passage through the digestive tract.

REPRODUCTION

Germ Cells and Fertilization

Rabbits do not show regular estrous cycles, that is, recurrent periods of sexual desire. During the breeding season the doe remains in heat for long periods of time. If she is not bred, the follicles in the ovary remain large and active for a period of 12 to 16 days. After this time they begin to regress. Meanwhile, new follicles grow to replace them. As a result, active follicles are present at all times during the breeding season. There may be a transitional period, while the new set of follicles is growing and the old set is retrogressing, when the doe lacks interest in the male and is temporarily sterile.

Ovulation usually does not occur in the female unless she is mated to a male. Following mating, the large follicles in the ovary begin to grow rapidly. These break about 10 hours after mating. This process of shedding the egg is called ovulation. In the meantime the sperm from the male move through the female tract to the upper part of the tubes so that when the eggs are shed from the follicles, the sperm enter and fertilize them. These fertilized eggs then undergo a number of changes and grow to become the developing fetuses. The young fetuses grow and develop in the uterus and this period of development usually takes 30 to 32 days. After the eggs are shed, the cells that line the follicles begin to grow and form small yellow bodies in the ovary, called the *corpora lutea*. These bodies secrete a hormone called progesterone, which is necessary to cause the uteri to grow and secrete substances that feed the developing young.

The practical method for measuring fertility in does is to determine the total number of live young born from each gestation. The litter size varies with the strain or breed, and the more fertile strains will produce an average of about eight young per pregnancy. Many factors, such as nutrition, heredity, and environment, affect fertility. Does that are underfed will not come in heat, and the quality of semen from starved bucks is lowered. Also, though experimental proof is lacking, there is a widespread opinion that does and bucks which are excessively fat have lowered reproductive capacities due to decreased sexual urge, or libido, or interference with the passage of eggs and sperm in the small reproductive tubules. Hereditary characters that affect fertility in does are number of eggs shed and fetal mortality. In highly fertile strains, the number of eggs shed averages about 10, while in low strains, the number may only be 4 or 5. In some low-fertility strains, normal numbers of ova are shed but an abnormally large number of embryos die during gestation. In fertile strains, about 15 to 20 percent of the fetuses die in the uterus during gestation. In some low-fertility strains, 80 percent of the fetuses die during the gestation period.

Gestation Period

The gestation period, or the period from mating to kindling, is 31 or 32 days. Some litters may be kindled as early as the 28th or 29th day, or as late as the 35th, but 98 percent of the normal litters will be kindled between the 30th and 33d day. If kindling is delayed 2 or 3 days, generally one or more of the fetuses is unusually large.

Age to Breed

The proper age of bucks and does for the first mating depends on breed and individual development. Smaller breeds develop more rapidly and are sexually mature at a much younger age than medium-weight or giant breeds. Does should be mated when they reach maturity; some difficulty may be experienced if mating is too long delayed. On the average, the smaller breeds may be bred when the bucks and does are 4 to

5 months old, the medium-weight breeds at 5 to 6 months, and the giant breeds at 8 to 10 months. Some individual rabbits within a breed develop more rapidly than others, and does usually mature earlier than bucks. In commercial production, it is the general practice to hold bucks a month longer than does before breeding for the first time, though there is no experimental proof that this is necessary.

Breeding Schedule

The breeding schedule you should follow is determined by the type of production. It probably would be better not to attempt to produce more than two or three litters a year in raising animals for show purposes. Arrange time of matings so that the offspring will be of proper age and development for the show classification desired. In commercial production for meat and fur, work breeding animals throughout the year if possible.

With a gestation period of 31 or 32 days and a nursing period of 8 weeks, a doe can produce four litters in a 12-month period if no failures or "passes" occur. Does of heavy producing strains can be mated 6 weeks after kindling, and, if no failures occur, will produce five litters in a year. Many commercial breeders are using breeding intervals of 21, 28, or 35 days after kindling to further increase the meat production of their herds. The general feeling is that for most efficient production, does should be worked to the extent of their genetic reproductive capacities. Experimental evidence is lacking as to what effect these rapid breeding schedules may have on the reproductive life of the doe, fryer development, feed conversion as measured by the pounds of feed necessary to produce a pound of meat, mortality, and carcass quality.

Where extreme temperatures make it undesirable to have litters kindled during 2 or 3 months of the year, does may be rebred 42 days after kindling and still produce four litters.

If a doe is full-fed a properly balanced ration during the suckling period, she should be in condition for breeding before the litter is weaned. If, however, the doe is not in good physical condition at the scheduled breeding time, she should not be bred until she is. If the litter is lost at kindling, or the size of the litter is materially reduced for other reasons, and the doe is in good condition, she may be rebred earlier than called for by the regular schedule, but not earlier than 3 or 4 days after kindling.

Lactation

During the last week of pregnancy the mammary glands develop rapidly. Though milk may be produced before kindling, and actually leak from the glands of high-producing does, the actual letdown and production is usually delayed until kindling, and is initiated under hormonal and nervous stimuli induced by the action of suckling. Maximum milk production is usually reached by the third week, after which production gradually declines. The duration of lactation varies depending upon diet, number of suckling young, and the length of time the young are left with the doe. Ordinarily, milk production is negligible after the sixth or seventh week, though in well-nourished, high-producing does with a litter of eight or nine, milk production may last for 8 weeks or longer. Milk has been observed in the stomachs of young weaned from the doe at 8 weeks of age, and milk can be expressed from the glands for several days thereafter.

The amount of milk produced depends upon several factors such as breed, strain, diet, and genetic constitution. Various studies of milk production in the rabbit indicate that during the height of lactation, milk yield may reach 35 grams per kilogram of live weight. On this basis, a 10-pound doe would produce approximately 140 grams (5 ounces) of milk per day.

Contrary to popular belief, the doe does not nurse her young throughout the 24-hour period. For the very young in the nest box, nursing is usually performed during the night or early morning. It may consist of a single feeding of only a few minutes. After the young leave the nest box and are consuming solid food they will try to nurse several times during the day. However, the doe will usually push them aside and restrict their nursing to the nighttime. Occasionally, does will allow the young to nurse during the day, as most rabbit breeders will agree. Students of animal behavior attribute the nursing habits of the doe to the fact that rabbits in their natural habitat are extensively preyed upon and rather helpless to defend their young. Therefore, it is advantageous for the doe to stay away from the young as much as possible.

Factors That Limit Conception

Among the causes of failure to conceive, or low conception rates, are false pregnancy (pseudopregnancy), season of the year, age, poor physical condition, sore hocks, injuries, and disease.

Pseudopregnancy.—Does may be mated or stimulated sexually and shed the egg cells, yet fail to become pregnant. This false pregnancy may be caused by an infertile mating or sexual excitement when one doe rides, or is ridden by, another. Does which become pseudopregnant are unable to conceive until the false-pregnancy period, which lasts 17 days, is over. After 18 to 22 days, the doe may give evidence of the termination of false pregnancy by pulling fur and attempting to make a nest. When false pregnancy has terminated, doe will resume normal reproductive activity and may be bred.

Separate does that are to be mated and put each in an individual hutch 18 days before mating. They will have passed through any false pregnancy period by mating time.

Season.—Spring is the optimal breeding season for the rabbit. The percentage of conceptions is higher at this time of year than at others.

Extreme temperatures, especially sudden changes to high temperatures, may cause the rabbits to go into a barren period that will continue for some time. Also, it is not unusual for the percentage of conceptions in a herd to show a marked decrease during the late summer and the fall. For example, at the U.S. Rabbit Experiment Station conception rates varied from a high of about 85 percent in March and April to a low of 50 percent or less in September and October. This is commonly referred to as the "fall breeding" problem in rabbits. The ovaries of the does may become inactive during the barren period, fail to produce normal egg cells, and occasionally shrivel. Where the bucks are not settling the does, the sperm cells may be inactive, low in vitality, abnormal, or absent.

Individual rabbits vary markedly as to duration of the barren period. Some does and bucks are fertile throughout the year for successive years. Others may go through periods of 4, 8, or 10 weeks when the does will not conceive or the bucks are sterile. Extreme cases in which no young are produced for 4 to 5 months may occur in herds where bucks and does are out of condition because the ration has been inadequate in quality or quantity, or both. If the herd has been properly cared for, most bucks and does should complete the barren period in 4 to 6 weeks.

Because does and bucks vary so much with respect to breeding, you may well consider this factor carefully. In selecting breeding stock, make your choice from offspring of parents that produce regularly.

Age.—Young does may not be sexually mature at the time of service, and old does may have passed their period of usefulness and fail to conceive. Do not attempt the first mating until the does are sexually mature and properly developed; the proper age is discussed on page 29.

Does should reproduce satisfactorily as long as they maintain good physical condition and properly nurse their litters. Retain them if younger and better stock is not available for replacements. In commercial herds, does that are properly cared for should produce litters until they are 2½ to 3 years old. An occasional individual rabbit may reproduce satisfactorily 4 to 6 years, or longer.

Physical Condition.—Rabbits that go "off feed," go into a prolonged or heavy molt, become abnormally fat or thin, or become out of condition for any reason, may have their reproductive powers impaired. The percentage that will conceive will be low, since they may become temporarily sterile.

Disease.—Never mate rabbits when they show any symptoms of disease. Remove such animals from the herd and hold them in quarantine until they recover.

Artificial Insemination

Artificial insemination has been practiced with rabbits for experimental purposes, but has not been applied to commercial breeding to any extent.

The semen from bucks averages about 0.5 cc. in volume, with a range of 0.1 to 6 cc. It contains about 700 million to 2 billion sperm per cubic centimeter. The total number of sperm per ejaculate averages 250 million, which does not mean a great deal, because of the extreme variation. The total number of sperm per ejaculate may range from 10 million to 12 billion.

Semen is collected from the bucks by means of an artificial vagina. After the artificial vagina has been prepared, the collection is made by using a doe for a mounting animal. The doe is taken to the buck's cage and when the buck mounts, the artificial vagina is placed between the buck and the doe. When the buck locates the artificial vagina, he will ejaculate into the open end with the same behavior as when breeding naturally. The operator must be alert to prevent the buck from breeding the doe. After the buck has been trained, a dummy made of a stuffed rabbit skin may be substituted for the mounting doe. If the ejaculate contains a clear gelatinous plug, it should be removed from the liquid portion of the semen.

A simple insemination tube has been described for insemination of the does. It consists of a glass tube and rubber bulb similar to a medicine dropper, with the last half-inch bent at a 30° angle. The bent end of the tube should have a diameter of ⅛-inch and the tube should be 3 to 3½ inches long. Droppers made from plastic tubes would be superior to glass because there is less danger from breakage. The ends of the tubes should be rounded to prevent damage to the vaginal walls.

The number of does bred to a buck depends on many factors, so an extreme range is possible. For an average buck, collections can be made at least twice weekly and the number of does bred will depend on the motility, density, and volume of the semen produced.

In general, artificial insemination is applicable to the rabbit industry, but not practical on a large-scale basis at the present time. Costs and technical problems associated with the establishment of studs, the processing and storage of semen, training and maintaining technicians, and scarcity of concentrated areas of production, tend to prohibit the use of artificial insemination as a practical tool of the industry. In addition, better means of evaluating bucks, and the development of more accurate selection indices, are necessary before artificial insemination in rabbits can be operated on a practical and profitable basis.

MANAGING THE HERD

Success in raising rabbits depends on efficient management. B e c o m e thoroughly acquainted with your animals—their characteristics and behavior, their likes and dislikes. Consideration for the welfare of animals is always necessary for success in raising them. Proper arrangement of equipment, hutches, and buildings is also essential to efficient management. When you e n t e r the rabbitry, do it quietly and make your presence known by speaking in a low tone. Caution others to do the same. Otherwise, the rabbits may become frightened, race around in the hutch and injure themselves, or jump into the nest boxes and injure the litters.

Methods of Handling Rabbits

Never lift rabbits by the ears or legs. Handling in this manner may injure them.

You can lift and comfortably carry small rabbits by grasping the loin region gently and firmly (fig. 12). Put the heel of the hand toward the tail of the animal. This method prevents bruising the carcass or damaging the pelt.

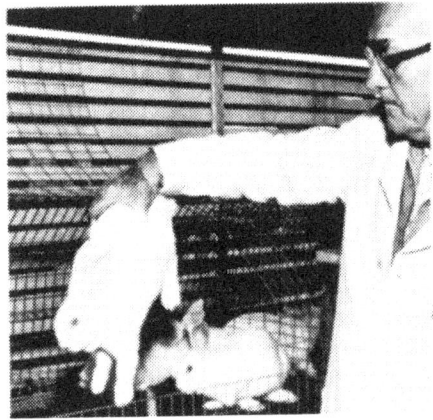

N 45957

FIGURE 12.—Proper way to carry small rabbits.

To lift and carry a medium-weight rabbit, let the right hand grasp the fold of skin over the rabbit's shoulder. Support the rabbit by placing the left hand under its rump (fig. 13).

N 45959

FIGURE 13.—Proper way to carry medium-sized and large rabbits.

Lift and carry heavier rabbits in a similar manner. If the rabbit scratches and struggles, hold it snugly under the left arm.

Making Matings

Does may give evidence of being ready for mating by restlessness, nervousness, efforts to join other rabbits in nearby hutches, and by rubbing their chin on feed mangers and water crocks. However, it is not necessary to depend on external signs to determine when a doe is to be bred. Set up a definite schedule and follow it, whether the doe shows signs of being ready for service or not.

Breed a large number of does at one time to make fryers available at a certain season for the trade, or utilize a breeding schedule to produce a constant supply. Always take the doe to the buck's hutch for service. You may have difficulty in service if you take the buck to the doe. The doe is likely to object to having another rabbit in her hutch and may savagely attack and injure the buck. Also,

some bucks are slow in performing service in a strange hutch. Mating should occur almost immediately on placing the doe in the buck's hutch. After the buck mounts and falls over on his side, the mating is accomplished. Return the doe to her own hutch.

It is difficult to get some does to accept service. Such does may be restrained for mating. To restrain the doe (fig. 14), use the right hand to hold the ears and a fold of the skin over the shoulders, and place the left hand under the body and between the hind legs. Place the thumb on the right side of the vulva, the index finger on the left side (you may prefer to use the index and second finger), and push the skin gently backward. This procedure throws the tail up over the back. Support the weight of the body by the left hand, and elevate the hindquarters only to the normal height for service.

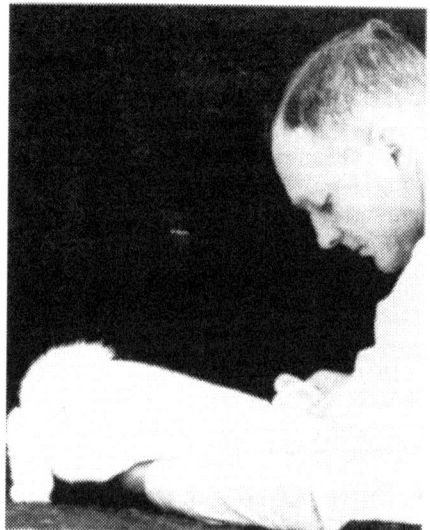

B 83066

FIGURE 14.—How to restrain a doe for mating when service is not promptly accepted. Shows position of hands for holding the doe and supporting and elevating the hindquarters.

COMMERCIAL RABBIT RAISING

Bucks and does accustomed to being handled will not object to such assistance. It is well to hold the doe in this way the first few times a young buck is used. This will expedite matings and insure service in difficult cases.

With a little patience and practice you can develop this technique to procure nearly 100-percent matings. This does not necessarily mean that all of the restrained will kindle, but the technique will help in increasing the number of kindlings.

Whether it is worthwhile to force-breed for increased pregnancies depends largely upon the number of does in production. Owners of large commercial rabbitries generally do not force-mate their does due to the increased labor and time involved. For small rabbitries a few extra litters could be worth the effort, and for breeders of pedigreed show stock, where animals are removed from production part of the year, forced breeding will help maintain a supply of replacements and stock for sale.

Maintain 1 buck for approximately each 10 breeding does. You can use mature, vigorous bucks several times a day for a short period.

Keep a breeding record showing date of mating and name or number of buck and doe.

Determining Pregnancy

It is not accurate to determine pregnancy by "test mating" (placing the doe in the buck's hutch periodically). Some does will accept service when pregnant and others will refuse service when they are not pregnant. Diagnosing pregnancy by noting the development of the abdominal region and gain in flesh is not dependable until late in pregnancy.

You can quickly and accurately determine pregnancy by palpating, after 12 to 14 days from mating, but you must handle the doe gently. The method for restraining the doe for palpating is illustrated in figure 15. The doe may be palpated in her own hutch or if it is more convenient she may be placed on a table covered with feed sacks or carpeting to prevent slipping. The ears and a fold of skin over the shoulders are held in the right or left hand; the other hand is placed under the shoulder between the hind legs and slightly in front of the pelvis; the thumb is placed on the right side and the fingers on the left side of the two uteri for palpating the fetuses. At 12 to 14 days following mating, the fetuses have developed into marble-shaped forms that are easy to distinguish as they slip between the thumb and fingers when the hand is gently moved forward and backward and a slight pressure is exerted (fig. 16). Caution must be used in this operation, because if too much pressure is exerted, the tissues may be bruised or torn loose from the walls of the uteri and a toxic condition or abortion may result.

There is less danger of bruising the tissues or causing the fetuses to be torn loose from the walls of the uteri in palpating at 12 to 14 days than at a later period. Also, diagnosing pregnancy after the 16th day of the gestation period is more complicated because the developing fetuses are so large that they may be confused with digestive organs. The inexperienced rabbit owner should make examinations at 12 to 14 days and then as he improves his technique and attains confidence in the operation, he may be able to develop the ability for diagnosing pregnancy accurately as early as the 7th or 8th

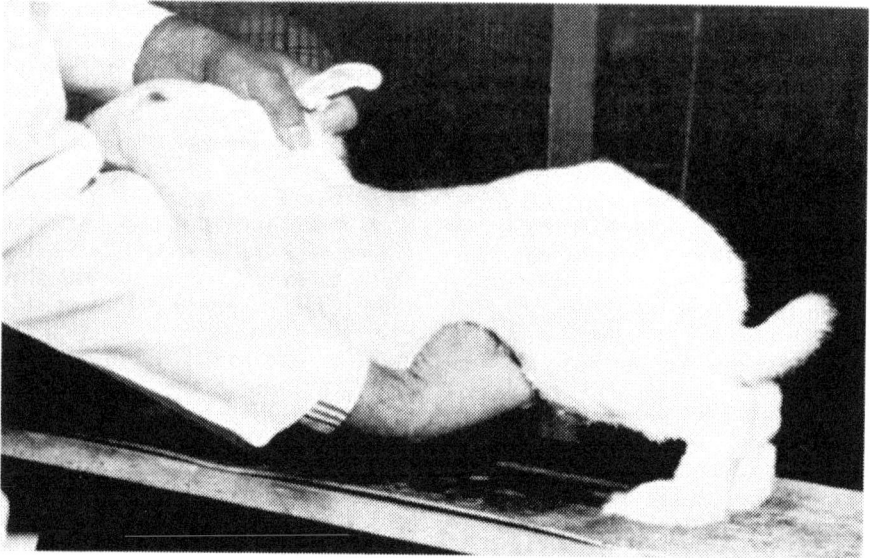

N 45951

FIGURE 15.—How to restrain a doe for palpating.

day. The chief advantage to be derived from palpating as early as the 7th or 8th day would be in the case of the breeder selling bred does. When it is desirable to ship bred does a considerable distance, diagnosing pregnancy at this early date makes it possible to have these does arrive at their destination in sufficient time to become settled and acquainted with their new environment, with the minimum risk of complications at kindling.

Figure 16 illustrates the continual increase in size of the uterus and the fetuses as pregnancy advances. The scale at the bottom of the illustration gives a means for arriving at a comparative estimate of the size of the fetuses. In each case, a fetus has been removed from the respective uterus. The 10-day fetus was so small that it does not show in the cut. By comparing the 14- and 21-day specimens it will be seen that the growth of the fetus is very rapid.

If, on palpating, no fetuses are found to be present, the doe has failed to conceive, in which case she should be rebred. The doe that is pregnant can be placed immediately on a diet that is best suited for pregnant does.

For the inexperienced person it would be good practice to repalpate a week later any does that have been diagnosed as nonpregnant. If a mistake has been made at the first handling the doe may then be given a nest box at the proper time before she is due to kindle.

Kindling

Place a nest box in the hutch about 27 days after the doe is mated. This allows the doe to prepare a nest in advance and assures a proper place for birth of the young.

Sometimes does fail to pull fur to cover their litter, or they kindle the litter on the hutch floor and let them become chilled. If you discover the young in time, you

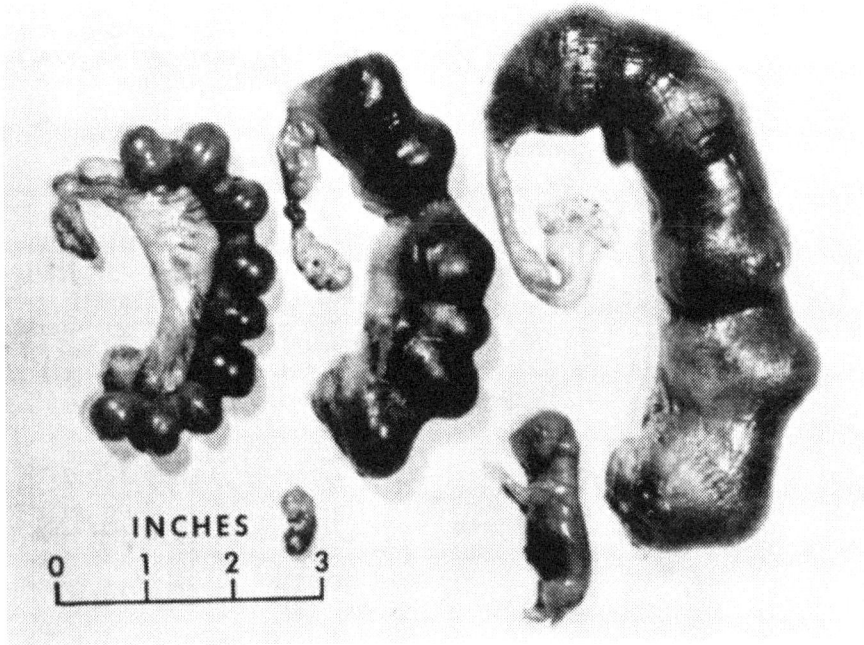

B 79008

FIGURE 16.—Uteri from three does showing embryonic development of 10-, 14-, and 21-day pregnancies. The 10-day embryo was so small that it did not show in the picture.

may be able to save them by warming, even if they appear to be lifeless. Arrange the bedding material to make a comfortable nest (fig. 17), and place the warmed young in it. The doe usually will take over from there. The doe's fur is easily removed at kindling time, and you can pull enough from the doe's body to cover the litter in the nest. It is advisable to keep extra fur on hand for such cases. Remove some fur from nests where does have pulled an excessive amount and keep it handy in a bag or box so it will remain clean. It is not necessary to sterilize or to deodorize the fur, or take any special measures to prevent the doe from smelling the strange fur.

A day or two before kindling, the doe usually consumes less food than normally. Do not disturb

her, but make her as comfortable as possible. You may tempt her at that time with small quantities of green feed. This will have a beneficial effect on her digestive system.

Most litters are kindled at night. After kindling, the doe may be restless. Do not disturb her until she has quieted down.

Complications at Kindling Time

Anterior, or breech presentation of young at birth is normal. If the doe is in proper condition for kindling, complications are rare. Pregnancy, however, makes a heavy demand on the doe and lowers her vitality, making her more susceptible to disease. A few days before or several days following kindling, pnuemonia may develop. If you are to treat pneu-

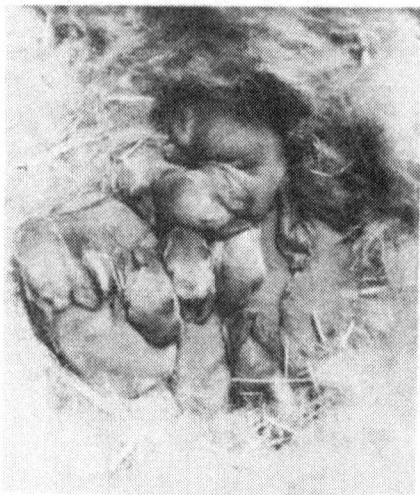

BN 21011

FIGURE 17.—Photo of new-born litter in nest.

monia successfully, you must detect it early. The doe's head is held high and tilted backwards. Breathing is difficult. Make the doe comfortable and add a little green feed to the ration if possible. Injections of penicillin or a penicillin-streptomycin mixture are effective in treating colds and pneumonia and reducing mortality.

Caked breast may be caused by the milk not being removed from the breast, or by injuries. Early symptoms of caked breast are firm, pink breasts that feel feverish to the touch. As caked breast develops, the tissues around the involved teats become enlarged and hard. The skin turns dark, the ends of the teats become discolored and tender, and the doe refuses to allow the young to nurse. Rub lanolin on the teats and massage the involved portion of the breast. You may restrain the doe to allow her own young or those from other litters to remove the milk. You may also strip the milk from the teats, taking care not to use too much pressure. *Do not lance the tissues.*

Mastitis, or "blue breast," is caused by bacterial infection and may be very contagious. The doe fails to consume her feed and is inactive. The breast is congested and feverish, turns red or purple, and the teats are discolored. Reduce feed, give some green feed, and inject penicillin intramuscularly in the thigh. (See treatment recommended for pneumonia.)

Care of Young Litter

On the day of kindling, or soon after, inspect the litter and remove any deformed, undersized, or dead young. If you are careful and quiet making the inspection, the doe generally will not object. *There is no danger of causing her to disown the young.* If she is nervous and irritable, place some tempting feed in the hutch immediately after inspection to distract her attention and quiet her.

Litters vary in size. The utility breeds usually average eight young. Some may number 12 to 18. For commercial purposes 7, 8, or 9 may be left with the doe. Does from strains that have been developed for heavy production may care for 9 or 10.

You can transfer some of the baby rabbits from a large litter to a foster mother that has a small litter. Adjusting the number of young to the capacity of the doe insures more uniform development and finish at weaning time. Mate several does so that they will kindle at about the same time. For best results, the young that are transferred should be within 3 or 4 days of the age of the foster mother's young.

Causes of Losses in Newborn Litters

If the doe is disturbed, she may kindle on the hutch floor and the litter may die from exposure. Even if predators—cats, snakes, rats, weasels, minks, bobcats, coyotes, strange dogs—cannot gain access to the rabbitry, they may be close enough for the doe to detect their presence, and she may be frightened and kindle prematurely. If she is disturbed after the litter is born and jumps into the nest box she may stamp with her back feet and injure or kill the newborn rabbits.

Occasionally a doe fails to produce milk. In such cases the young will starve within 2 or 3 days unless the condition is noted and the young transferred to foster mothers. Keep a close check on newborn litters for several days after birth to make sure they are being fed and cared for properly.

Does sometimes eat their young. This may result from a ration inadequate in either quantity or quality, or from the nervousness of a doe disturbed after kindling. It is also possible that the doe is of a strain that exhibits poor maternal instincts. Does usually do not kill and eat healthy young, but limit their cannibalism to young born dead, or those that are injured and have died. Proper feeding and handling during pregnancy will do more than anything else to prevent this tendency. Give another chance to a valuable doe that destroys her first litter; if she continues the practice, dispose of her.

Weaning

Under most management programs the young are weaned at 8 weeks of age. At that age young meat rabbits should average 4 pounds in weight and be ready for market. Some commercial producers leave the young with the doe for 9 or 10 weeks to get a 4½- to 5½-pound fryer. Small litters (fewer than five young) can be weaned at an earlier age and the doe rebred. Also, under accelerated breeding programs where does are bred less than 35 days following kindling, it is advisable to wean the young at 5, 6, or 7 weeks of age to allow the doe to prepare for her next litter. It is best to allow a few days between removal of one litter and birth of the next. For example: if a doe is bred 28 days after kindling, it is possible to leave the litter with her until they are 56 days of age, allowing for kindling about 3 days later. It depends upon the condition of the doe and her ability to stand up under this type of program. You may wish to remove the young at 7 weeks of age and give the doe 7 to 10 days to prepare for the next kindling.

Determining the Sex of Young Rabbits

Separate the sexes at weaning, if you are saving junior replacements, or breeding stock. It is possible to determine accurately the sex of baby rabbits less than a week of age, but it is easier to do so when they are weaned. To keep the rabbit from struggling, restrain it firmly, yet gently. A commonly used method is to hold the rabbit on its back between your legs with the head up. With your left hand restrain the rabbit around the chest holding the front legs forward alongside the head. Using the right hand, place the thumb behind the right hind leg and use the index and forefinger to depress the tail backward and downward. The thumb is then used to gently depress the area in front of the sex organs to expose the reddish mucous membrane. In the buck, the organ will protrude as a rounded tip, while in the doe the membrane will protrude to form a slit with a depression at the end next to the anus.

Marking for Identification

Mark each breeding rabbit for identification. Tattooing the ears is a satisfactory method. When properly done, it is permanent and will not disfigure the ears. You can obtain instruments for the purpose from biological and livestock supply houses. Ear tags and clips are not satisfactory because they tear out and disfigure the ear. Identification is then lost. An adjustable box is convenient for restraining the rabbits for tattooing (fig. 18). With this equipment, one person can do the job.

Castration

Castration of bucks may be desirable; for example, where An-goras are to be kept in colonies for wool production. In producing domestic rabbit meat for market, there are no advantages to be derived from castrating bucks for improving the rate of growth and condition, reducing the quantity of feed required to produce a pound of gain in live weight, and improving the carcass and pelt. Probably the only advantage to be derived would be that it reduces fighting and makes possible the maintaining of a number of castrated bucks in one inclosure, thereby saving equipment, time, and labor. Castration is a simple operation, most easily performed when bucks are 3 to 4 months old. You also can perform it at weaning time.

To restrain an animal for the operation, have an assistant hold the buck's left forefoot and left hind foot with his left hand, and the right forefoot and right hind foot with his right hand, with the animal's back held firmly, but gently, against his lap. Clip all the wool from the scrotum. Disinfect a

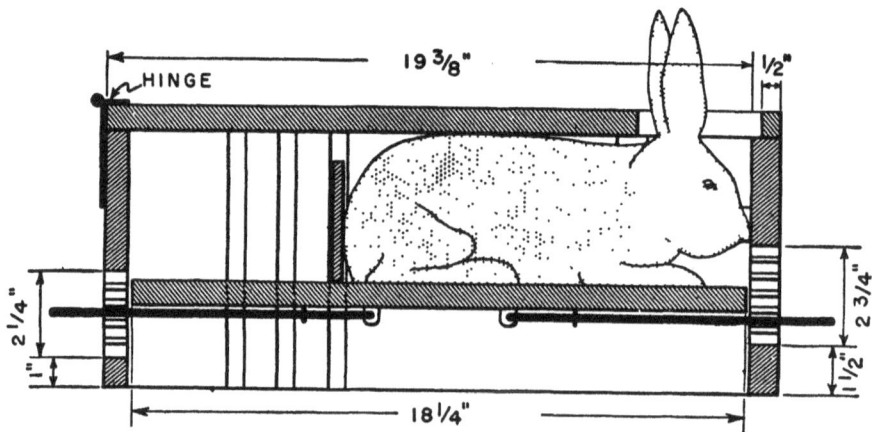

FIGURE 18.—Vertical section of a box for restraining a rabbit for tattooing. The spring-type holders tacked to the lower side of a movable floor compress the rabbit toward the top of the box. A movable cross partition holds the rabbit toward the front. Blocks of wood on each side hold the rabbit's head in the center of the hole at the top.

sharp knife or razor blade. If you do not use a disinfecting agent on the rabbit, he will lick the wound frequently, keeping it clean and the tissues soft, thus promoting healing.

Press one of the testicles out into the scrotum. Hold it firmly between the thumb and forefinger of the left hand. Make an incision parallel to the median line and well toward the back end of the scrotum to allow the wound to drain readily. To keep the testicle from being drawn up into the abdominal cavity, as soon as it comes from the incision pull it out far enough from the body for the cord to be severed just above it. To prevent excessive hemorrhage, sever the cord by scraping with a knife rather than by cutting. If too much tension is put on the cord and it is drawn too far from the body, injury may be brought about by internal hemorrhage or other complication.

After the second testicle has been removed in the same manner, lift the scrotum to make sure that the ends of the cord go back into the cavity.

Handle the animal gently. After the operation, place it in a clean hutch where it can be quiet and comfortable.

Care of Herd During Extreme Temperatures

Heat.—In almost all sections of the United States high summer temperatures necessitate some changes in the general care and management of rabbits. Provide adequate shade to the animals during the hottest part of the day. Good circulation of air throughout the rabbitry is necessary, but avoid strong drafts. Provide an abundant supply of water at all times.

Newborn litters and does advanced in pregnancy are most susceptible to high temperatures. Heat suffering in the young is characterized by exteme restlessness; in older animals, by rapid respiration, excessive m o i s t u r e around the mouth, and occasionally slight hemorrhages around the nostrils. Move rabbits that show symptoms of suffering from the heat to a quiet, well-ventilated place. Give them a feed sack moistened with cold water to lie on. Water crocks and large bottles filled with cracked ice and placed in the hutch so that the rabbits can lie next to them contribute to the rabbit's comfort.

In well-ventilated rabbitries, wetting the tops of the hutches and the floors of the houses on a hot, dry day will reduce the temperatures 6 to 10 degrees F. The tops of hutches should be waterproof, as rabbits should be kept dry. You can use overhead sprinkling equipment in houses with concrete or soil floors that drain readily or sprinklers above the roof of rabbit sheds (fig. 2). You can install a thermostatically controlled sprinkler that will work automatically.

The use of evaporative coolers on the roof or sides of the buildings, whereby air is drawn over wet pads and distributed through the building, can be successfully used in hot, dry climates such as the southwest United States. This type of cooler is widely used in homes and can be adapted to use in rabbitries which are partially, or totally, enclosed.

In areas of high humidity, the use of sprinklers or extra water will aggravate the situation and add to the rabbit's discomfort. Under such conditions, it is advisable to install fans, or place the buildings to take advantage of all breezes, in order to get maximum movement of air. The use of refrigerant air conditioning is usually uneconomical and impractical, due to the high initial investment and operating costs.

During the summer it sometimes is difficult to regulate the quantity of fur in the nest box to keep the litter comfortable. A cooling basket (fig. 19) then will provide relief for the young. It is useful from the time the young are kindled until their eyes are opened and they are able to care for themselves. Make this basket 15 inches long, 6 inches wide, and 6 inches deep. Use 1/8-inch mesh hardware cloth 15 by 18 inches; two boards 3/4 by 6 by 6 inches; two laths 3/8 by 1 1/2 by 15 inches; and 2 1/2-inch screwhooks. Tack the hardware cloth to three edges of the two square boards. To keep the basket from bending, nail the laths lengthwise, in front and back of the basket outside the wire. Nail the top edges of the laths flush with the tops of the end boards. At the back, insert two screwhooks in the

end boards about 2 inches from the top, so you can hang up the basket. When the temperature is high enough to make the young restless, place them in the basket. Hang up the basket inside the hutch near the top and leave it for the day. In the evening, if it is cooler, return the litter to the nest box. Where high temperatures continue throughout the night, place the young in the nest box for a short time in the evening for nursing. Replace them in the basket for the night and allow them to nurse again in the morning.

Do not hang the basket in direct sunlight.

Cold.—Mature rabbits, if kept out of drafts, suffer little from low temperatures. However, precautions should be taken to protect rabbits from direct exposure to rain, sleet, snow, and winds. If they are enclosed in a building, care must be taken to provide adequate ventilation and to prevent the

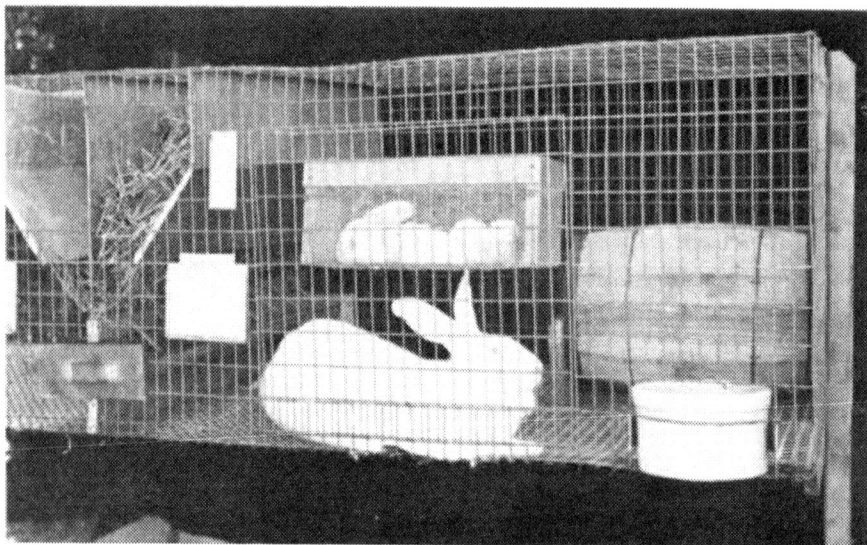

83068-B

FIGURE 19.—A cooling basket hung in the hutch to provide comfort for the young during hot weather.

accumulation of moisture. Cold weather, drafts, and high humidity are conducive to the spread of respiratory infections. For young litters, provide nest boxes and sufficient bedding to keep them warm as discussed previously.

Controlled Environments.—The use of controlled environment in rabbitries, where rabbits are maintained under more or less constant environmental conditions, is receiving increased attention. Several large commercial rabbitries in the western United States are changing to or are constructing this type of housing. The advantages of controlled environment are the elimination of extremes in weather and, perhaps, seasonal fluctuations in production.

Preventing Injuries

Paralyzed hindquarters in rabbits usually result from improper handling or from injuries caused by slipping in the hutch while exercising or attempting to escape predators, especially around kindling time. Such slipping usually occurs at night. Common injuries are dislocated vertebras, damaged nerve tissue, or strained muscles or tendons. If the injury is mild, the animal may recover in a few days. Make the injured animal comfortable and feed it a balanced diet. If it does not improve within a week, destroy it to prevent unnecessary suffering. It is important, therefore, that your rabbits be provided with quiet, comfortable surroundings and be protected from predators and unnecessary disturbances.

The toenails of rabbits confined in hutches do not wear normally. They may even become long enough to cause foot deformity. The nails may also catch in the wire mesh floor and cause injury and suffering. Periodically cut the nails with side cutting pliers. Cut below the tip of the cone in the toenail. The cone can be observed by holding the foot up to daylight. This will not cause hemorrhaging or injury to the sensitive portion.

Preventing Sore Dewlaps

During warm weather the dewlap, or fold of skin under the rabbit's chin, may become sore. This is caused by drinking frequently from crocks and keeping the fur on the dewlap wet so long that it becomes foul and turns green. The skin on the dewlap and on the inside of the front legs becomes rough and the fur may be shed. The animal scratches the irritated area, causing abrasions and infection.

Remove the cause by placing a board or brick under the water crock to raise it so that the dewlap will not get wet when the rabbit drinks. If the skin becomes infected, clip off the fur and treat the area with a medicated ointment until the irritation clears up. The best solution to the problem is to use an automatic dewdrop watering system which eliminates the possibility of wet dewlaps.

Sanitation and Disease Control

To protect the herd's health, keep the rabbitry equipment sanitary. Remove manure and soiled bedding at frequent intervals and contaminated feed daily. Inspect water crocks and feed troughs daily and wash them frequently in hot, soapy water. Rinse them in clear water, allow them to drain well, and place them in the direct rays of the sun to dry. If it is impractical to sun the equipment properly, rinse it first in water to which a disinfectant has been added and then in clear water.

To prevent or control a disease

or parasitic infection, thoroughly disinfect hutches and equipment which have been occupied, or used, by sick animals, or where excessive mortality has occurred. One of the coal tar derivatives or household disinfectants may be used. Allow hutches and equipment to dry before returning rabbits to the hutches. A large blowtorch or weedburner may be used periodically to remove hair and cobwebs and to disinfect the hutches. Clean and disinfect nest boxes before using them a second time.

Maintaining sanitary conditions in the rabbitry is a preventive measure for controlling disease in the herd. Be constantly on the alert for the appearance of any sign that might indicate disease. Isolate animals suspected of having disease at least 2 weeks to determine definitely whether they are dangerous to the health of the herd. Place newly acquired rabbits and those returned from shows in quarantine at least 2 weeks for the same reason. Burn or bury dead animals.

Using hutches with self-cleaning floors, guards on feed troughs, and feed hoppers, will aid greatly in internal parasite control by protecting feed from contamination.

The most serious disease of domestic rabbits is pasteurellosis. This disease manifests itself in a wide variety of conditions such as pneumonia, snuffles (sinusitis), and other respiratory infections; and septicemia, a generalized blood infection.

Another serious problem in rabbit health is enteritis, or bloat. Three types of enteritis are distinguished: diarrhea, mucoid, and hemorrhagic. The specific cause of enteritis is not known and there are no reliable measures for prevention or treatment.

Coccidiosis, both of the liver and intestines, is a serious problem in some areas but can be successfully treated.

The tapeworms which infest the rabbit are those which at a later stage infest dogs and cats, but the rabbits seem to suffer little harm from them.

Tularemia, the disease that has in recent years killed off so many wild rabbits, is spread by ticks and fleas. If domestic rabbits are kept in clean conditions, free from ticks and fleas, they will not contract it.

Domestic rabbits suffer from other ailments such as fungal infections, mange, sore hocks, and spirochetosis or vent disease, but these usually can be successfully treated and do not present a major problem.

These and other ailments of domestic rabbits are described in table 4.

TABLE 4.—*Common ailments of domestic rabbits*

Diseases and symptoms	Cause	Treatment and control
Ear Mange or Canker: Shaking of head, scratching of ears. Brown scaly crusts at base of inner ear.	Ear mites (*Psoroptes cuniculi* (rabbit and goat ear mite) and *Notoedres cati* (cat ear mite).)	Into each ear, pour 1 oz. of a 5 percent lime-sulfur solution (prepared by mixing commercial 30 percent lime-sulfur concentrate, 1 part, water 5 parts). Massage solution over inner and outer surfaces of ears; repeat as necessary. Rubber gloves advisable.

TABLE 4.—*Common ailments of domestic rabbits*—Continued

Diseases and symptoms	Cause	Treatment and control
Skin Mange: Reddened, scaly skin, intense itching and scratching, some loss of fur.	Mites (*Cheyletiella parasitivorax* (rabbit fur mite) and *Sarcoptes scabiei* (scabies or itch mite).)	Dip entire animal in a 1.75 percent lime-sulfur bath (prepared by mixing commercial 30 percent lime-sulfur concentrate, 8 oz., laundry detergent, 1 tablespoonful, per gallon tepid water). Repeat in 2 weeks if necessary. Rubber gloves advisable.
Favus or Ringworm: Circular patches of scaly skin with red, elevated crusts. Usually starts on head. Fur may break off or fall out.	Fungus (*Trichophyton*, and *Microsporum*).	Griseofulvin given orally at the rate of 10 milligrams per pound body weight for 14 days. Combine this treatment with dusting nest boxes with industrial fungicidal sulfur. Can also be treated with a brand of hexetidine. Apply to infected area for 7 to 14 days.
Sore Hocks· Bruised, infected, or abscessed areas on hocks. May be found on front feet in severe cases. Animal shifts weight to front feet to help hocks.	Bruised or chafed areas become infected. Caused by wet floors, irritation from wire or nervous "stompers."	Small lesions may be helped by placing animal on lath platform or on ground. Advanced cases are best culled. Medication is temporarily effective.
Urine-Hutch Burn: Inflammation of external sex organs and anus. Area may form crusts and bleed and, if severely infected, pus will be produced.	Bacterial infection of the membranes.	Keep hutch floors clean and dry. Pay particular attention to corners where animals urinate. Daily applications of lanolin may be of benefit.
Spirochetosis or Vent Disease: Similar lesions as produced by urine or hutch burn. Raw lesions or scabs appear on sex organs; transmitted by mating.	Spirochete (*Treponema cuniculi*).	Inject intramuscularly 100,000 units of penicillin. Do not breed until lesions heal. If only a few animals infected, it is easier to cull than treat. Do not loan bucks.
Conjunctivitis or Weepy Eye: Inflammation of the eyelids; discharge may be thin and watery or thick and purulent. Fur around the eye may become wet and matted.	Bacterial infection of the eyelids; also may be due to irritation from smoke, dust, sprays, or fumes.	Early cases may be cleared up with eye ointments, argyrol, yellow oxide of mercury, or antibiotic. A combination of 400,000 units of penicillin combined with ½ gr. streptomycin to each 2 ml. For eye infections drop directly into eye. Protect animals from airborne irritants.

TABLE 4.—*Common ailments of domestic rabbits*—Continued

Diseases and symptoms	Cause	Treatment and control
Caked Breasts: Breasts become firm and congested, later hard knots form at sides of nipples. Knots may break open, showing dried milk.	Milk not drawn from glands as fast as formed, because of too few young, or young not nursing sufficiently; usually a management problem with high milk-producing does.	Do not wean young abruptly; if litter is lost, rebreed doe and protect doe from disturbance so young can nurse properly. Correct faulty nest boxes that injure breasts.
Mastitis or Blue Breasts: Breasts become feverish and pink, nipples red and dark. Temperature above normal, appetite poor, breasts turn black and purplish.	Bacterial infection of the breasts (*Staphylococcus* or *Streptococcus*).	Inject 100,000 units of penicillin intramuscularly twice each day for 3 to 5 days. Disinfect hutch and reduce feed concentrates. If severe case, destroy. NEVER transfer young from infected doe to another doe.
Snuffles or Cold· Sneezing, rubbing nose; nasal discharge may be thick or thin. Mats fur on inside of front feet. May develop into pneumonia, usually chronic type of infection.	Bacterial infection of the nasal sinuses (*Pasteurella multocida* or *Bordetella bronchiseptica*).	Individual animals may be treated with a combination of 400,000 units of penicillin combined with ½ gr. streptomycin to each 2 ml. Give intramuscularly 1 ml. for fryer size, 2 ml. for mature Repeat on 3d day.
Pneumonia Labored breathing with nose held high, bluish color to eyes and ears. Lungs show congestion, red, mottled, moist, may be filled with pus. Often secondary to enteritis.	Bacterial infection of the lungs. Organisms involved may be *Pasteurella multocida*, *Bordetella bronchiseptica*, and *Staphylococcus* and *Streptococcus* sp.	If the above treatment is started early, it is effective. For control in herds, add feed grade sulfaquinoxaline so that level will be 0.025 percent, feed 3 to 4 weeks. Water soluble sulfaquinoxaline can be added at a level of 0.025 percent and fed 2 to 3 weeks
Heat Prostration: Rapid respiration, prostration, blood-tinged fluid from nose and mouth Does that are due to kindle are most susceptible.	Extreme outside temperature. Degree varies with location and humidity.	Reduce temperature with water sprays, foggers. Place wet burlap in hutch or wet the animal to help reduce body temperature.
Coccidiosis, Intestinal: Mild cases, no symptoms; moderate cases, diarrhea and no weight gain. Severe cases have pot belly, diarrhea with mucus, and pneumonia is often secondary.	Parasitic infection of the intestinal tract caused by coccidia. (*Eimeria perforans, E. magna, E. media, E. irrisidua.*)	Keep floor clean, dry, remove droppings frequently. Prevent fecal contamination of feed and water. Add feed grade sulfaquinoxaline so that level will be 0.025 percent, feed 3 to 4 weeks. Water soluble sulfaquinoxaline can be added at level of 0.025

TABLE 4.—*Common ailments of domestic rabbits*—Continued

Diseases and symptoms	Cause	Treatment and control
		percent and fed 2 to 3 weeks. These treatments, combined with sanitation, will greatly reduce numbers of parasites and animals infected.
Enteritis, Bloat, or Scours: Loss of appetite, little activity, eyes dull and squinted, fur rough, and animals may appear bloated. Diarrhea or mucus droppings; animals may grind teeth. Stomach contents fluid, gaseous, or filled with mucus.	Unknown; never has been shown to be infectious or transmitted to other animals.	Add 50 gr. furazoladine per ton of feed to give final concentration of 0.0055 percent. Feed intermittently or continuously. Water soluble chlortetracycline or oxytetracycline at a level of 1 pound to 100–150 gal. of water may be used for treating individual cases; too costly for herd control.
Fur Block: Animals reduce feed intake or stop eating completely, fur becomes rough, and weight is lost. Stomach filled with undigested fur, blocking passage to intestinal tract. Pneumonia may become secondary.	Lack of sufficient fiber, bulk, or roughage in the diet. Junior does or developing does most susceptible.	Increase fiber or roughage in the ration. Feed dry alfalfa or timothy hay.
Tapeworm Larvae: White streaks in liver or small white cysts attached to membrane on stomach or intestines Usually cannot detect in live animals.	Larval stage of the dog tapeworms (*Taenia pisiformis*) or of the cat tapeworm (*T. taeniaeformis*).	No treatment; keep dogs and cats away from feed, water, and nest box material Eggs of tapeworm occur in droppings of dogs and cats.
Pinworms: No specific symptoms in live animals. White threadlike worms found in cecum and large intestine cause slight local irritation.	Pinworms (*Passalurus ambiquus*).	None; infection not considered one of economic importance.
Metritis or White Discharge: White sticky discharge from female organs, often confused with sediment in urine. Enlarged uterus detected on palpation. One or both uteri filled with white, purulent material.	Infection of the uterus by a variety of bacteria, nonspecific.	Dispose of infected animals and disinfect hutches. Infected area difficult to medicate. When both uteri are infected, animal is sterile.

TABLE 4.—*Common ailments of domestic rabbits*—Continued

Diseases and symptoms	Cause	Treatment and control
Pasteurellosis: May be an acute or chronic infection. Nasal discharge, watery eyes, weight loss, or mortality without symptoms. Inflammation and consolidation of lungs, inflammation of bronchi and nasal sinuses.	Bacterial infection (*Pasteurella multocida*).	Individual animals may be treated with a combination of 400,000 units of penicillin combined with ½ gr. streptomycin to each 2 ml. Give intramuscularly 1 ml. for fryer size, 2 ml. for mature. Repeat on 3d day. For herd control, add feed grade sulfaquinoxaline at level of 0.025 percent, feed 3 to 4 weeks. Save replacement stock from clean animals and cull out chronically infected animals. Use good sanitary measures to reduce transmission to new animals.
Paralyzed Hindquarters: Found in mature does, hind legs drag, cannot support weight of pelvis or stand. Urinary bladder fills but does not empty.	Injury, resulting in broken back, displaced disc, damage to spinal cord or nerves.	Protect animals from disturbing factors, predators, night prowlers, and visitors or noises that startle animals, especially pregnant does.
Wry Neck· Head twisted to one side, animals roll over, cannot maintain equilibrium	Infection of the organs of balance in the inner ear. May be parasitic or bacterial.	None, eliminate ear canker from herd. Some cases result from nest-box injuries.

Effective treatments are not known for many rabbit diseases. It is usually simpler and safer to destroy a few sick animals than it is to treat them and risk spreading infection to healthy stock. This is especially true of animals with respiratory infections.

Fur-Eating Habit

Rabbits that eat their own fur or bedding material, or gnaw the fur on other rabbits, usually do so because the diet is inadequate in quality or quantity. A common cause is a diet low in fiber or bulk. Sometimes the protein content of the diet is too low. Adding more soybean, peanut, sesame, or linseed meal may correct the deficiency.

The experienced breeder notes the condition of each animal in the herd and regulates the quantity of feed to meet its individual requirement. Providing good-quality hay or feeding fresh, sound green feed or root crops as a supplement to the grain or pelleted diet also helps to correct an abnormal appetite.

Preventing Fur Block

In cleaning themselves by licking their coats, or when eating fur from other animals, rabbits swallow some wool or fur which is not digested. The only noticeable result may be droppings fastened together by fur fibers. However, if the rabbit swallows any appreciable amount, it may collect in the stomach and form a "fur block" that interferes with digestion. If it becomes large enough, it blocks the alimentary tract and the animal starves. The most satisfactory method of preventing this is to shear Angoras regularly, and try to prevent fur eating among your rabbits by providing adequate roughage and protein in their diet. A block of wood or other material upon which the rabbit can chew may be used to reduce fur chewing.

Gnawing Wooden Parts of the Hutch

Gnawing wood is natural for the rabbit. Protect wooden parts of the hutch by placing wire mesh on the inside of the frame when constructing the hutch and by using strips of tin to protect exposed wooden edges. Treating the wood with creosote protects it as long as the scent and taste last. Placing twigs or pieces of soft wood in the hutch protect it to some extent; rabbits may chew these instead of the hutch.

Rabbits that have access to good-quality hay and are receiving some fresh green feed or root crops are less likely to gnaw on their hutches.

Disposal of Rabbit Manure

Rabbit manure has a high nitrogen content when the rabbits are fed a well-balanced diet (*3*). It will not burn lawns or plants and is easy to incorporate in the soil. It is satisfactory on gardens and lawns and about flowering plants,

shrubbery, and trees. There is no danger in using it for fertilizing soil on which crops are to be raised for feeding rabbits.

The value of rabbit manure depends on how it is cared for and used. There will be less loss of fertilizing elements if the material is immediately incorporated into the soil. When manure is stored in piles and exposed to the weather, chemicals are lost through leaching and heat. Much of this loss can be prevented by keeping the manure in a compost heap or in a bin or pit.

Earthworms in the Rabbitry

Where earthworms are active throughout the year as in warm climates, they may be used to advantage under rabbit hutches to save labor in removing fertilizer. Make bins for confining the worms the same length and width as the hutch and 1 foot deep (fig. 20). Place bins on the ground, not on solid floors, and keep the fertilizer moist to insure the worms working throughout the bin.

Earthworms convert the rabbit droppings into casts—a convenient form of fertilizer for use with flowers, lawns, shrubs, trees, and other foliage. If you keep a large population of worms, there will be no objectionable odor. Very few flies will breed in the bins. It is necessary to remove the manure only at 5- to 6-month intervals.

Records and Recordkeeping

A convenient and simple system of records is essential for keeping track of breeding, kindling, and weaning operations. You can use the information in culling unproductive animals and in selecting breeding stock. The essential features of a simple record system are illustrated in the hutch card and the buck breeding record card shown in figures 21 and 22.

The USDA does not furnish record cards. They may be obtained from firms dealing in supplies for the rabbitry or you may prepare your own. Some feed mills also furnish their customers with hutch cards and record forms.

N 45949

FIGURE 20.—Worm bins installed beneath rabbit hutches.

HUTCH CARD

Animal No. __W 301__ Born __12/12/55__ Breed __New Zealand White__
Sire __W 394-__ Dam __W 604__ Litter No. __W 714__

DATE BRED	BUCK NO.	DATE KINDLED	NUMBER YOUNG BORN ALIVE	DEAD	NUMBER YOUNG RETAINED	LITTER NO.	DATE WEANED	NUMBER WEANED
6/1/56	W418-	7/2	11	0	8	W19	8/27	8
8/24/56	W418-	9/24	9	0	8	W175	11/19	8
11/16/56	W418-	Passed	11/30					
11/30/56	W421-	12/30	9	1	8	W316	2/24/57	8
2/21/57	W421-	3/24	11	0	8	W465	5/19	7
Ⓐ								

PRODUCTION RECORD

LITTER NO.	WEANING NUMBER	AGE	WEIGHT	NOTES:
W19	8	56	30.2	
W175	8	56	31.0	
Passed	11/30			
W316	8	56	32.0	
W465	7	56	28.0	
Ⓑ				

FIGURE 21.—Hutch card, a useful form of record. A, front; B, back.

BUCK BREEDING RECORD

Buck No._____

Breed _____   Sire _____

Date born _____   Dam _____

| Doe | Location | Date Bred | Result of breeding | | | | Weaned | |
| | | | Kindled | | Passed | | | |
			Alive	Dead	Date		Number	Weight

FIGURE 22.—Sample of buck breeding record.

TYPES OF PRODUCTION

If you want a fair income from your commercial herd, you must be able to care for a large number of rabbits. Your returns will be in direct ratio to the number and quality of does maintained and your efficiency of management.

Formerly, about 10 man-hours each year were required to care for a doe and her four litters. With improved hutch and feeding equipment (figs. 23 and 24), rations designed to save labor in feeding, palpation of does, and other herd management practices, the number of man-hours has been greatly reduced. It is now possible for a breeder to care for more than twice as many does in the same length of time with less effort.

Fryer Production

According to the regulations governing the grading and inspection of domestic rabbits, issued by the Department of Agriculture, "A fryer or young rabbit is a young domestic rabbit carcass weighing not less than 1½ pounds and rarely more than 3½ pounds processed from a rabbit usually less than 12 weeks of age," (14, Title 7, Part 54, Section 54.261).

Rabbits raised for meat and fur usually are marketed when they

N 45952

FIGURE 23.—Filling feed cart from bulk feeder tank.

N 45947

FIGURE 24.—Filling hopper (self-feeder) from feed cart.

reach fryer weight even though the pelts are not prime. In order to yield a carcass weighing from 1½ to 3½ pounds, young rabbits should have a live weight of approximately 3 to 6 pounds. Best carcass yields are usually from young rabbits weighing from 4 to 4¾ pounds, when weaned at 2 months of age (figs. 25 and 26). These should yield a carcass (including liver and heart) of 50 to 59 percent of the live weight, 78 to 80 percent of which is edible.

For fryer production, medium-weight to heavyweight breeds are preferred. Their young are most apt to develop to the desired weight and finish by the time they are 2 months old.

A pound of marketable fryer rabbit will require 2½ to 3½ pounds of feed, or a total of approximately 100 pounds for a doe and litter of 8, from mating of the doe to marketing of the young when 2 months old. Good does nurse their litters 6 to 8 weeks. The young develop more rapidly if they are in the hutch with their mothers until they are 8 weeks of age. By that time, the milk supply will have decreased, the young will be accustomed to consuming other feed, and weaning will be less of a shock than if undertaken at an earlier age. Young that are weaned and held for several days before market may either fail to gain or actually lose weight. Therefore, it is usually best to leave the young with their mothers until they go to market.

If you want to produce fryers heavier than those weaned when 56 days old, keep young rabbits with their mother an additional 8 or 9 days. These fryers should gain an average of 0.6 to 1.0 pound during this period, if full fed a balanced diet. However, they will require more pounds of feed per pound of

N 45953

FIGURE 25.—Young meat rabbits waiting shipment to market.

BN 26085

FIGURE 26.—Litter of fryer rabbits at market age and weight.

increase in live weight than previously, and the death of one rabbit in a litter during the extra holding period may eliminate any additional profit.

Roaster Production

According to the regulations governing the grading and inspection of domestic rabbits, issued by the Department of Agriculture (14, Title 7, Part 54, section 54.-262), "A roaster or mature rabbit is a mature or old domestic rabbit carcass of any weight, but usually over 4 pounds processed from a rabbit usually 8 months of age or older."

You can fatten culls from the breeding herd for roasters, if they are in good condition. In some areas you may find it profitable to develop young rabbits to heavier weights primarily for the meat market. Such rabbits should yield a carcass that is 55 to 65 percent of the live weight, with 87 to 90 percent of it edible. However, the quantity of feed required to produce a pound of gain, live weight, increases with each pound of gain, and may amount to 12 to 14 pounds to increase the live weight from 9 to 10 pounds. Therefore, the cost of feed required to produce these gains must be assessed against the value of the heavier rabbits. Unless a premium is paid for mature rabbits for their meat or better fur quality, it is doubtful if such production would be more profitable than that of rabbits of fryer weight.

Castrated bucks require less time and about 5 percent less feed than normal bucks to attain a given live weight. As indicated previously, one advantage of castration is that a number of animals can be kept together with a saving of equipment, time, and labor. If a buck is castrated when 2 months old, his skin at maturity will grade as a doe skin and sell for a higher price. These factors, however, usually do not justify the extra work and danger involved in castration.

ANGORA RABBIT WOOL PRODUCTION

Angora rabbits are raised primarily for wool production (fig. 27). Wool on Angoras grows to a length of 2½ to 3½ inches each 3 months, or approximately 1 inch per month. You can shear 14 to 15 ounces of wool a year from a mature Angora that is not nursing young. This wool is valued for its softness, warmth, and strength. It is used in blends with other fibers in the manufacture of children's clothing, sport clothes, garment trimmings, and clothes for general wear. Used alone it is usually too light and fluffy, and blends create better tensile strength and durability.

There are two main types of Angora rabbits—the English and the French. Present standards of the American Rabbit Breeders Association, however, make English and French types of wool synonymous. It is difficult to distinguish the English Angora rabbits from the French when they are off type, and the choice largely is a matter of personal preference. The typical French Angora usually is larger than the English. The wool fiber of the French is shorter and coarser than that of the English, but wool yield is greater. Owing to competition with other fibers, both natural and synthetic, and competition with imported Angora rabbit wool, the market price is generally low and it is advisable to use the Angora as a dual purpose animal for both meat and wool production. The commercial Angora weighs at least 8 pounds and is being bred more and more to improve its quality for meat.

Keep herd bucks and does in individual hutches. Keep woollers—does and castrated bucks maintained primarily for wool production—in groups or colonies to

11514-D

FIGURE 27.—Angora rabbit.

save labor. Castration of bucks that are to be reserved for wool production may reduce fighting in the herd, but wool production is not increased by castration.

To prevent infestation with internal parasites and to keep wool clean, install self-cleaning floors in the pens.

Wool should be harvested prior to breeding to prevent mauling, and soiling of the wool.

Angoras are generally sheared or plucked every 10 to 11 weeks though some producers pluck their animals monthly and some at intervals beyond 3 months.

Feed and care for Angoras in the same way as for other breeds. Because of their long wool, however, you must handle Angoras to find out how much flesh they are carrying; determine the amount of flesh by running the hand along the backbone. Reduce or increase the quantity of feed to keep the animals in condition.

A properly constructed manger for hay and green feed, or the use of a hopper, protects the wool from foreign matter and prevents contamination of feed.

Equipment for Grooming and Shearing

You will need the following equipment for grooming and shearing:

A table, waist-high, with a 12- by 24-inch top covered with carpet or a feed sack to keep the rabbit from slipping, and equipped with castors to allow easy turning. A table equipped with straps or cords for restraining the rabbit is advantageous.

A hairbrush with single steel bristles set in rubber, for brushing and removing foreign material from wool.

A pair of barber's scissors or electric clippers.

A ruler for measuring the length of wool.

Containers for storing wool.

Grooming.—Commercial woollers require little, if any, grooming between shearings provided they are properly cared for and sheared every 10 to 12 weeks. If you allow the coat to grow for a longer period, the fibers may become webbed, tangled, or matted.

For grooming, place the rabbit on the table. Part the wool down the middle of the back. Brush one side, stroking downward. As you reach the end of the wool, brush upward and outward to remove all foreign material. Make another part in the wool about half an inch farther down the side. Repeat the operation until the job is completed. Groom the other side the same way.

For grooming the head, front legs, and belly place the rabbit on its back in your lap. Hold its hindquarters gently but firmly between the knees. Separate small areas of wool and groom the way you did the sides.

For grooming the hind legs, place the rabbit on its back in your lap. Hold the head and front feet under the left arm. Use the left hand to hold the rabbit's hind legs.

Shearing.—Before shearing, cut off all stained ends of wool. Place the back of the scissors against the rabbit's body to prevent cutting the skin. Begin at the rump and shear a strip about an inch wide to the neck. Repeat this operation until you have removed all the wool

from one side. Turn the rabbit around and repeat the shearing operation on the other side, starting at the neck and shearing toward the rump. For shearing the head, front legs, belly, and hind legs, restrain the rabbit as for grooming. Separate small areas of wool and shear the way you did the sides. Do not injure the doe's teats. Do not shear wool from the belly of a pregnant doe. After shearing, lightly brush the rabbit to straighten out the wool fibers and prevent the formation of mats.

During cold weather, newly sheared rabbits need protection. A nest box in the hutch affords adequate protection during cool spells. When the temperature is as low as 30° to 40° F., keep the animals in a building where you can maintain comfortable temperatures. In winter, leave a half inch of wool on the body for protection.

Grading, Preparing, and Marketing Wool

Label a container for each grade of wool and place it near the shearing table. Grade the wool as sheared. Following are the usual commercial grades:

Plucked wool:	
Super _____	3¼ inches or longer
No. 1 _____	3 inches or longer
No. 2 _____	2 inches or longer
Sheared wool:	
No. 1 _____	2¼ to 3 inches
No. 2 _____	1½ to 2 inches
No. 3 _____	1 to 1½ inches
Shorts _____	½ to ¾ inches (may be slightly webbed)
No. 4 _____	Matted
No. 5 _____	Stained and unclean

While the above have been the usual accepted commercial grades for Angora rabbit wool, some grading systems have been simplified to the extent that only three grades are considered: No.

1—clean wool; No. 2—clean mats; and No. 3—all other wool, including soiled wool.

Put each grade in a separate paper bag without packing too tightly. Tie the bags and place them in burlap sacks or corrugated boxes for shipment.

If the wool is to be stored, place it in a tightly covered container. To protect the wool from moths, put mothballs or crystals in a small sack before placing this in the container with the wool.

Some Angora breeders spin the wool on an old-fashioned spinning wheel and knit the yarn into garments for home use or for sale. Others sell wool to organizations or individuals who collect large quantities and these organizations or individuals in turn sell to mills.

MARKETING

Slaughtering and Skinning

Slaughter in clean, sanitary quarters. Obtain information on regulations and restrictions from local health authorities.

The preferred method of slaughtering a rabbit is by dislocating the neck. Hold the animal by its hind legs with the left hand. Place the thumb of the right hand on the neck just back of the ears, with the four fingers extended under the chin (fig. 28). Push down on the neck with the right hand, stretching the animal. Press down with the thumb. Then raise the animal's head by a quick movement and dislocate the neck. The animal becomes unconscious and ceases struggling. This method is instantaneous and painless when done correctly.

Another method is to hold the animal with one hand at the small of the back, with its head down, and stun it by a heavy blow at the base of the skull.

Suspend the carcass on a hook inserted between the tendon and the bone of the right hind leg just above the hock (fig. 29). Remove the head immediately to permit thorough bleeding so the meat will have a good color. Remove the tail and the free rear leg at the hock joint, and cut off the front

feet. Then cut the skin just below the hock of the suspended right leg and open it on the inside of the leg to the root of the tail, continuing the incision to the hock of

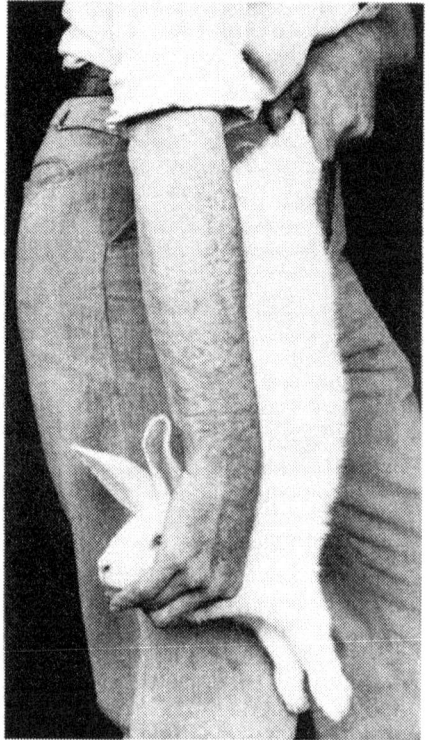

83078 B

FIGURE 28.—How to hold a rabbit for dislocating neck in slaughtering.

83079-B

FIGURE 29.—Steps in skinning rabbits and removing internal organs. Small jets of water from pipe beneath rack wash blood from back panel and trough.

the left leg. Carefully separate the edges of the skin from the carcass, taking special pains to leave all fat on the carcass as the skin is pulled down over the animal. This makes a more attractive meat product, facilities drying the skin, and prevents "fat burns" on the pelt in drying.

Even small cuts lessen the value of a skin. As soon as you remove the skin, place it on a stretcher, secure it and hang it up for drying. (See section on "Rabbitskins.")

After skinning the carcass, make a slit along the median line of the belly and remove the entrails and gall bladder. Leave the liver and kidneys in place. Remove the right hind foot by severing at the hock. Take particular care not to get hairs on the carcass; they are difficult to remove, detract from the appearance, and are unsani-

tary. Rinsing the carcass in cold water facilitates removal of hair and blood and also cleans the carcass. Brush the rabbit's neck thoroughly in water to remove any blood. Do not leave the carcass in water more than 30 minutes; prolonged soaking causes it to absorb water, and water in the meat is adulteration.

Chill the carcass in a refrigerated cooler. Arrange the carcass on a cooling rack so that moderate air movements and a suitable temperature within the cooler will reduce the internal temperature of the carcass to no less than 36° F. and to no more than 40° within 24 hours.

Hanging by the hind legs for chilling may cause a carcass to be drawn out of shape, so that the pieces will not fit satisfactorily into a carton. Some processors chill carcasses in wire trays, arranging them so the pieces will be of a proper shape for packaging.

Cutting and Packaging Rabbit Meat

Hotels, restaurants, hospitals, clubs, and other establishments usually purchase the whole carcass. Their chefs prefer to cut them to meet their own requirements. Housewives usually prefer the cut-up, packaged product. Cut up the fryer rabbit with a knife; using a cleaver may splinter the bones. Common cuts from fryer carcasses are illustrated in figures 30 and 31. In large commercial processing plants, a bandsaw is used. A paraffined box with a cellophane window makes a neat, sanitary package for the chilled rabbit carcass (fig. 31). If the package is to be handled considerably or the meat is to be frozen, use a box without the cellophane window, but wrap the meat or the box in a special salable wrapping to prevent freezer burns and loss in palatibility.

A box 9 inches long, 4 inches wide, and 2½ inches deep is suitable for a fryer carcass weighing 1¾ to 2¼ pounds. Arrange the cuts attractively. Include the heart, kidneys, and liver.

If you sell to the home trade or furnish butchers with meat that is to be consumed locally, you can make a neat, sanitary, and inexpensive package by arranging the pieces of fryer and a sprig of parsley on a paper plate and covering them with a piece of clear cellophane or other wrapping material (fig. 31).

For information on regulations governing the grading and inspection of domestic rabbits and specifications for classes, standards, and grades, write to the Consumer and Marketing Service, U.S. Department of Agriculture, Washington, D.C., 20250.

Crating and Shipping Live Rabbits

You can ship rabbits almost any distance with safety, if they are in good condition, properly crated, and provided with food and water. Do not ship them in extremely hot or cold weather. Always use well-ventilated crates that are long enough to permit the rabbit to lie down. Use straw, not sawdust, for bedding. Crates with slanting tops discourage stacking (fig. 32).

Put only one animal in a compartment of a shipping crate. Animals to be in transit 24 hours or less need only a small quantity of feed and water. If the trip is long, more feed and water are needed. It is wise to attach to each crate a bag of feed and a printed request to feed and water the animals once daily. Plenty of fresh water and feed should be accessible to the rabbits at all time. For rabbits in transit use the type of feed given in the rabbitry. As an alternative, a bunch of fresh carrots placed in the crate will provide enough feed and moisture for several days' travel, and eliminates the possibility of spilling feed and water supplied in containers.

Label the crate clearly, advising against exposing the animals to sun or rain, and also against placing the crates near steam pipes. Notify the purchaser when rabbits are shipped.

You can make shipping crates from packing boxes. It is good business, however, and effective advertising, to ship rabbits in durable crates that are neatly built, light in weight, and attractive. Furnish ample space in each compartment and see to it that wire netting keeps the rabbits from gnawing the wood.

BN 26086

FIGURE 30.—Some common cuts obtained from a rabbit carcass.

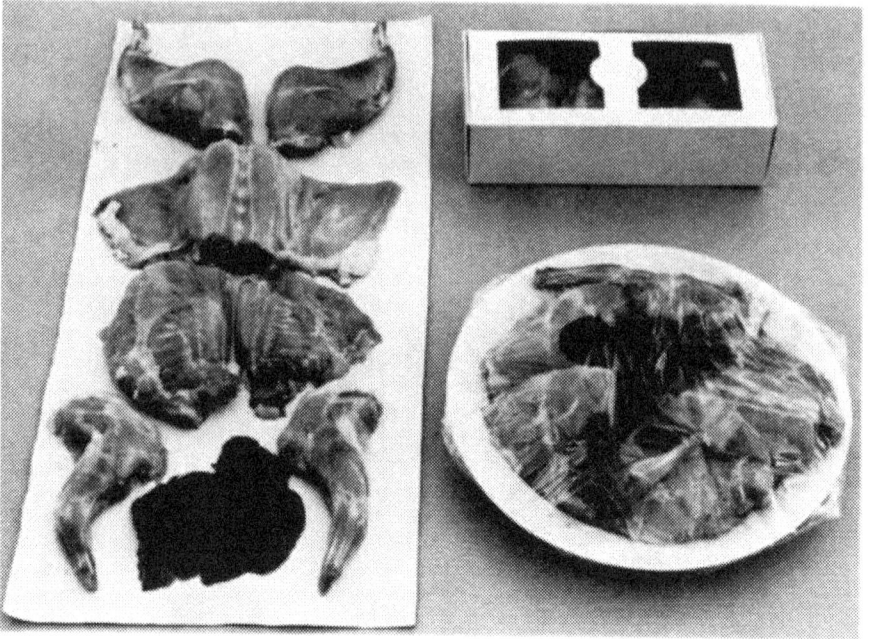

83080 B

FIGURE 31.—Some common cuts obtained from a rabbit carcass, and samples of preparation for sale.

BN 26083

FIGURE 32.—Homemade shipping crates for transporting rabbits.

RABBITSKINS

Curing

While still warm, place skins to be cured flesh side out on wire or board formers or shapers (with the fore part over the narrow end). Take care to remove all wrinkles. You can make a satisfactory skin shaper from 5 feet of No. 9 galvanized wire. This equipment has been called a "stretcher," but the term may give a wrong impression. It is not desirable to stretch the skin unduly. Mount a skin on the shaper, making sure both front feet casings are on the same side, and fasten it with clothespins (fig. 33). This arrangement lessens in-jury to the fur of the back, which is the most valuable. On the day after skinning, examine the pelts to see that the edges are drying flat, that the skin of the front legs is straightened out, and that any patches of fat are removed.

All skins must be thoroughly dried before you pack them, but do not dry them in the sun or by artificial heat. Hang them up so the air can circulate freely about them. If you will not ship the dried skins for some time, hang them in loose bundles of 50 in a cool, dry place away from rats and mice. In the summer or in a warm climate, sprinkle the stored skins with

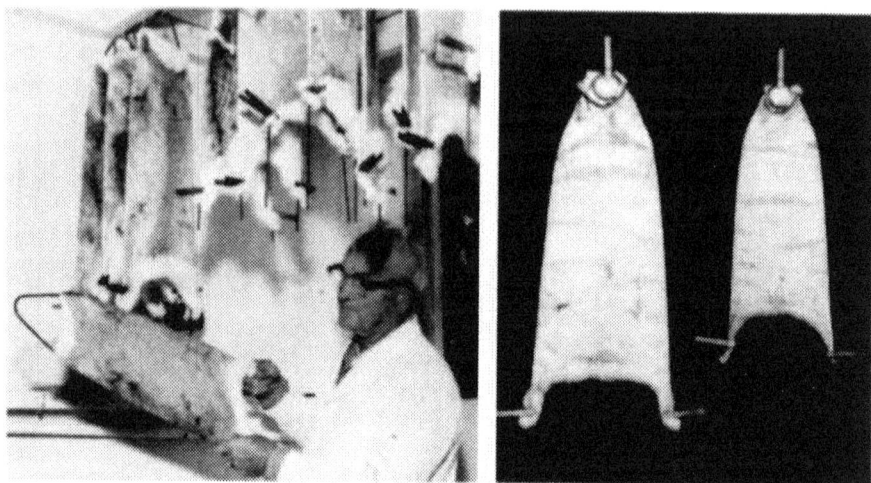

FIGURE 33.—How to place a rabbit pelt on a shaper before hanging it up to dry (left). Two sizes of rabbit pelts properly placed on wire shapers (right). Front feet casings (not shown in picture) are on other side of shaper.

naphtha flakes. *Never use salt in curing rabbitskins.*

Marketing

Domestic r a b b i t s k i n s vary greatly in density and quality, depending on the degree of care that breeders take in breeding. Good fur can be produced on efficient meat-producing animals by selective mating. Better skins command higher prices.

Because of the relative cheapness of rabbitskins, volume is necessary for the dealer to market them satisfactorily; and since d r e s s i n g charges are so much per skin, the larger skins, other things being equal, will bring the better price even when they are sold by the pound. Whether it will pay to grow or condition heavier rabbits for the market depends on the relative cost of feed and the market value of the finished product. In areas where similar skins are produced in quantity, it might be profitable for several rabbit raisers to market their skins cooperatively.

White skins bring higher prices than colored skins because of the adaptability to use in the lighter shades of garments and hats.

If good and poor skins of different sizes and colors are mixed in a shipment, the entire shipment is usually accepted at the price of poor skins. Sort the skins (unless you have too few) and offer them in separate shipments.

Grades

All rabbitskins have some value in the fur trade. About 85 percent of domestic rabbitskins are from rabbits 8 to 10 weeks old. These skins are known in the trade as "fryer skins." They are usually sold by the pound as butcher run, that is, ungraded. Five or six fryer skins usually weigh a pound. In full-fed rabbits weighing 4 to 12 pounds, the poorest skins come from animals up to 134 days old. Older animals produce a higher percentage of better grade skins. The better grade skins from older domestic rabbits are usually sold by the piece, primarily because they are larger than fryer skins.

Raw-fur buyers usually grade rabbitskins as firsts, seconds, thirds, and hatters. Many buyers have their special grades. Firsts and seconds may be divided into five color classes—white, red, blue, chinchilla, and mixed. Some furriers also grade firsts and seconds as large, medium, and small. If you have enough skins, pack white, red, blue, and chinchilla skins separately by colors. Put skins of all other colors together.

Firsts are prime pelts that are large, properly shaped, and properly dried. All the hair is held firmly in the skin and the skin side is free from fat, flesh, pigment spots, streaks, and cuts. The thicker and denser the underfur, the more valuable the pelt and the better price it will bring.

Firsts are used for making garments. They may be sheared or used in the natural or "long-haired" condition. They also may be used in the natural color or may be dyed. A uniform, dense underfur is necessary to make desirable rabbitskin garments. The coarse, longer hairs should return to their natural position and present a smooth appearance immediately after the hand is passed through and against the natural flow of the hair coat.

Fryer skins contain only a small percentage of fur usuable for garments primarily because of shedding or molting marks and secondarily because of thin fur and leather. Rabbitskins for fur garments have been in bad repute because inferior grades were used in the past.

Seconds are pelts that have shorter fur and less underfur than firsts. The unprime colored skin shows dark pigment spots or streaks and, sometimes, large black splotches on the leather side. These markings do not show on white skins since pigment is lacking. Seconds also include pelts that are improperly shaped and dried, have been damaged in shipment, or show poor spots where the skin has been pierced or the fur is short or missing.

Thirds are pelts with short fur and thin underfur and those from animals too young or those that are shedding. Thirds are of no value to the furriers. They are used in the manufacture of toys, specialty articles, and felt for hats.

All skins that do not meet requirements of the other grades are "hatters." Pelts that are badly cut or otherwise mutilated, or poorly stretched and dried, also are classed as hatters. The underfur of such pelts is used in making felt. Since the denser skins yield more cut fur, the hat trade pays more for them.

The distribution of domestic rabbitskins into these several grades depends on the demand for each kind. The market may be such that practically all the rabbitskins at a given time will be sold as "hatters." Under some conditions, there may be but little demand even in the hat trade.

Packing and Shipping

To avoid spoilage or damage in transit, take care in packing skins for shipment.

So far as possible keep skins in the same shape as when removed from the form. Carefully examine each one to make sure that it is properly dried. Do not pack or ship a moist pelt or one that has patches of oily fat on it. Make up large quantities of skins into bales. Sprinkle naphthalene or paradichlorobenzene on every two or three layers of skins, as you pack them. This will keep out insects that might cause damage. When a bale has been made up, cover with burlap, sew with strong cord or binder twine, and mark. Always protect skins when ship-ping them. Ship smaller quantities in gunny or feed sacks. Do not use wooden boxes for shipping rabbitskins; the weight adds materially to shipping charges.

ECONOMICS OF RABBIT PRODUCTION

The commercial rabbitry is one that is operated for the "profitable" production of rabbit meat. A great deal of progress has been made in the development of the rabbit industry. It is a stable enterprise for an individual employing sound management principles, but numbers alone do not spell success. The rabbitry, as any business, cannot afford additional units of production unless those added provide returns equal to, or greater than, their costs. The determination of the point at which marginal costs of production equal marginal returns necessitates the presence of two factors important for the success of a commercial rabbitry: Concise, accurate and current records, and close attention to care and feeding of the rabbits.

Records

Records need not be extremely detailed, unless the personal desires and time of the operator allow for minute recordkeeping. Whatever records are kept should permit the operator to calculate costs of production and evaluate the progress made over comparable periods of time.

Information basically desirable is (1) the number of does bred, (2) the number of conceptions, (3) the number of does kindling, (4) the number of does raising a litter, (5) total young left with doe, and (6) total number of young weaned or raised per breeding. These facts will provide the necessary permanent production factors. Information can be obtained from the hutch record cards (fig. 21) and accumulated daily on a monthly summary form. The monthly figures can then be accumulated on an annual summary form, and an annual summary of the rabbitry can be ascertained by posting the accumulated investment, income, and expense figures on a summary chart.

Labor

Close attention to the rabbits is essential to success. Although it is not practical to provide care 24 hours a day, too little labor is disastrous. Somewhere between no care and constant care is an optimum amount of labor for a rabbitry of a given size.

Rabbitry management studies conducted in San Bernardino County, Calif., in 1962 and 1963 revealed a range of 2.3 to 27.7 hours of labor per doe per year, with an average of 6.4 hours per doe. Rabbitry sizes, based on the average number of working does, ranged from a low of 31.4 to a high of 792.9, with an average of 247.5. For all practical purposes, the labor source reported by participating rabbitries was a husband-and-wife team. Using the averages of the management studies conducted, an average daily requirement of 6.4 hours per doe per year for a 247.5-doe rabbitry gives a total annual labor need of 1,584 hours. This gives conjecture to the possibility of operating a rabbitry on a 5-day week. A weekly work schedule of 40 hours totals 2,080 hours per year. Dividing 2,080 hours by the study average of 6.4 hours of labor per doe gives a total of 325 does required to use the hours of labor available. This is 77.5 more than the management study average.

Certainly, a progressive industry will keep in tune with the innovations continuously appearing on the market. These available engineering ingenuities can be used to advantage by the rabbit industry. Where needed, modifications can be made for their adoption. Self-feeders, automatic watering devices, built-in sunken nest boxes, and electric carts for feeding and carrying fryers or breeding stock are but examples. Constant striving for improved feeds and feeding techniques, building designs, and materials and for stock improvement through breeding provides the progressive rabbitry operator with the means to reduce the hours of labor per animal.

Naturally, these innovations will tend to increase investment costs of the rabbitry. However, if these innovations reduce the marginal cost of production, there will be justification for their use.

Investment

The rabbit management studies conducted in San Bernardino County, Calif., during the years 1962 and 1963 showed the following investment costs *per doe:* [4]

Investment	1962	1963	2-year average
Land	$12.80	$10.07	$11.48
Building and equipment	17.09	18.06	17.57
Miscellaneous supplies	.01	.03	.02
Feed	.61	.48	.55
Stock	5.48	5.87	5.67
Total	36.07	34.51	35.29

[4] Although these figures are for 1962 and 1963, they can be converted to present costs.

The reduction in land investment from 1962 to 1963 does not indicate lowering land values, but rather that rabbitries participating in the study in 1963 were located on lower valued land than those participating in 1962.

Costs for housing and equipment vary considerably, but the 1962-63 quotation for an all-wire hutch indicated the retail charge was $5.07 per hole or per doe. Accessory equipment, such as nest boxes, feeders, medication tank, feed tank, etc., cost $4.11, for a total cost of $9.18 per doe. Deducting this amount from the 2-year average building and equipment charge of $17.57, as shown in the tabulation, leaves a cost of $8.39 per doe for the building and its related electrical and watering equipment. The above costs reflect commercial retail prices for new equipment. Reductions can be made by canvassing the market for used material available. The use of either semirigid or flexible plastic pipe in lieu of galvanized pipe for water lines should be explored. Frequently, when salvage material is adequate, it can substantially reduce building costs. Bargaining ability can further reduce costs, but reductions should not be made at the expense of reliable, sound, and safe construction.

Returns and Expenses

The rabbit industry represents a relatively stable market. In the management studies conducted in 1962 and 1963, the price received per pound of fryer for the 2-year period ranged only 4 cents, and that received per pound of mature rabbits ranged only 2 cents. In each case, the changes were gradual and infrequent. Total receipts reported for the participating rabbitries averaged $29.59 per doe per year. On the average, sale of fryers represented 90 percent of total receipts.

The study averages of 6.4 hours of labor and of $29.59 income per doe give a gross income return of $4.62 per hour.

The management study for 1962 reported that total production costs (feed, labor, stock purchased, miscellaneous expense, depreciation, and interest at 6 percent on investment) amounted to $22.54, for an average of 104.2 pounds of meat produced per doe. The costs reported for 1963 totaled $23.55, for the 115.9 pounds of meat produced per doe. Production costs for the 2 years average $23.04 per doe, or $3.60 per hour, based on 6.4 hours labor per doe. Subtracting this from a gross income return of $4.62 per hour leaves $1.02 per hour as net income. The average cost per pound of meat sold in 1962 amounted to 21.6 cents, and in 1963, 20.0 cents.

The following tabulation presents a summary of the percentage distribution of costs in the San Bernadino County management studies for 1962 and 1963:

	Percent
Feed	53.8
Labor	30.5
Miscellaneous	5.6
Depreciation	4.5
Interest	5.6

The distribution of income, by source, was as follows:

	Percent
Fryer rabbits sold	89.9
Miscellaneous income	1.7
Breeding stock sold	4.0
Mature stock sold	2.7
Inventory change	1.7

Although it is to be remembered that conditions vary from year to year and from area to area, as well as from rabbitry to rabbitry, these summary figures were substantiated by those of a 1964 survey conducted in the Santa Clara-Alameda County area of northern California.

LITERATURE CITED

(1) AMERICAN RABBIT BREEDERS ASSO-
CIATION, INC.
1960. STANDARD OF PERFECTION.
111 pp., illus. Pittsburgh,
Pa.

(2) CASSADY, R. B.
1962. MALOCCLUSION, OR "BUCK
TEETH," IN RABBITS. U.S.
Dept. Agr., Agr. Res. Serv.
CA–44–48. 3 pp., illus.

(3) ———
1962. VALUE AND USE OF RABBIT
MANURE. U.S. Dept. Agr.,
Agr. Res. Serv. CA–44–47.
5 pp.

(4) ———
1963. RABBIT MEAT IS COMPETI-
TIVE. U.S. Dept. Agr., Agr.
Res. Serv. CA–44–38. 2 pp.

(5) ———
1965. THE "FALL BREEDING"
PROBLEM IN RABBITS. U.S.
Dept. Agr., Agr. Res. Serv.
CA–44–34. (A rev. edition.)
2 pp.

(6) HAGEN, K. W., JR.
1962. TULAREMIA, AN ANIMAL-
BORNE DISEASE. U.S. Dept.
Agr., Agr. Res. Serv. CA–
44–49. 3 pp.

(7) ——— and LUND, E. E.
1964. COMMON DISEASES OF DO-
MESTIC RABBITS. U.S. Dept.
Agr. ARS–45–3–2 (Rev.)
8 pp.

(8) HARDY, T. M. P., and DOLNICK,
E. H.
1948. ANGORA RABBIT WOOL PRO-
DUCTION. U.S. Dept. Agr.
Cir. No. 785. 22 pp., illus.

(9) HINER, R. L.
1962. PHYSICAL COMPOSITION OF
FRYER RABBITS OF PRIME,

CHOICE, AND COMMERCIAL
GRADES. U.S. Dept. Agr.,
Agr. Res. Serv. CA–44–37.
6 pp.

(10) MORRISON, F. B.
1956. FEEDS AND FEEDING. 22d
ed. 1165 pp., illus. Ithaca,
N.Y.

(11) SANDFORD, J. C.
1957. THE DOMESTIC RABBIT. 258
pp., illus. London.

(12) SMITH, S. E., DONEFER, E., and
CASADY, R. B.
1966. NUTRIENT REQUIREMENTS
OF RABBITS. Natl. Acad.
Sci.—Natl. Res. Council
Pub. 1194, 17 pp., illus.

(13) TEMPLETON, G. S.
1962. DOMESTIC RABBIT PRODUC-
TION. 3d ed. 213 pp., illus.
Illinois.

(14) UNITED STATES CONSUMER AND
MARKETING SERVICE.
1967. REGULATIONS GOVERNING
THE GRADING AND INSPEC-
TION OF DOMESTIC RABBITS
AND EDIBLE PRODUCTS
THEREOF AND UNITED
STATES SPECIFICATIONS FOR
CLASSES, STANDARDS, AND
GRADES WITH RESPECT
THERETO. Title 7 Code of
Federal Regulations, pt. 54,
effective August 1966.

(15) UNITED STATES DEPARTMENT OF
AGRICULTURE.
1962. INHERITANCE OF "WOOLLY"
IN RABBIT. U.S. Dept.
Agr., Agr. Res. Serv. CA–
44–36. 3 pp.

(16) ———
1964. RAISING RABBITS. U.S.
Dept. Agr. Farmers' Bul.
2131. 24 pp., illus.

www.ingramcontent.com/pod-product-compliance
Lightning Source LLC
Chambersburg PA
CBHW031814190326

41518CB00006B/335